Status Report
German Society of Refrigeration and Air Conditioning, DKV
No. 23

Edmund Altenkirch

Pioneer of Refrigeration and Heat Transformation

2023

German Society of Refrigeration and Air Conditioning, DKV
Striehlstrasse 11, DE 30159 Hannover, Germany

Translator's Note

As I learned in my early childhood, Edmund Altenkirch was the first employer of my father, Siegfried Unger. In those times—the early 1950s—my family and I lived in the "Koblenzer Straße" in Neuenhagen near Berlin. Altenkirch's property was near our residence, about one kilometre away, a mansion which belonged to a former member of the ruling party of the former German Reich. Altenkirch lived near the city train station that connected Neuenhagen by a three-quarter hour electric train ride to the city center of Berlin and beyond that the boundary of West Berlin.

My father, in those days a student at the Humboldt University in Berlin, worked in his free time for Altenkirch and was fascinated by his research because Altenkirch was preoccupied with improving the Absorption Refrigeration Machine, which was an antecedent of the absorption refrigerator which was itself the precursor of the later compression refrigerator, which was distributed throughout the world. I remember seeing the Absorption Refrigeration Machine, in which circulated a pink fluid in intertwined glass pipes and mirrored flasks (cf. Figure 6.3 and 6.5, page 75 and 81). The fluid was almost soundless as it circulated, making only a slight bubbling that you could hear.

As I learned from this book, the gas bubble transport of the fluid by carrier gas drove the solution cycle and the fluid was propelled by the concentration gradient. Mechanical moving parts were omitted, and capillarity was utilized to maintain the machine going. It made a big visual impact, seeming almost like a perpetual motion machine! After the death of Altenkirch, my father took over the leadership of his laboratory, located in the research establishment on his property, and had this machine which as a child I admired rebuilt.

My father was fascinated by this type of machine. His later wish, when he was no longer living in Neuenhagen but in Berlin, was to develop it further, but he became preoccupied with other scientific subjects.

Another project Altenkirch pursued was something we would today call "green technology": a tropical dwelling with solar cooling (see Figure 7.6, page 92).

Altenkirch had a big property with a large orchard and four buildings. The one on the western side of the plot was his own residential living area, a mansion with a tower room where one could retreat. In this building he lived with his wife Margarete, who carried out all the administrative work for him. There was a second building where later his second daughter, Eva, with her husband and two sons lived, as well as an office building where the laboratory was also located. In a fourth building on the eastern side of the property lived his youngest child, his son Joachim.

I also knew Altenkirch's eldest daughter Erika. A photograph taken on the occasion of his birthday shows Altenkirch with his whole family around him. Around that period that I knew him and his family—the early 1950s—Erika and her immediate family emigrated from East Germany to Canada, where she pursued her own research into the relationship between science, human values, and the future.

Michael Unger, Berlin, February 2023

Edmund Altenkirch—Pioneer of Refrigeration and Heat Transformation

Subtitle:
Reversible Thermodynamic Processes in Refrigeration, Heat-Pumping, and Air-Conditioning

Author
Prof. Dr. Dr. Siegfried Unger
graduate physicist

Co-author
Jörn Schwarz
graduate engineer

Transcription from the German of the authors into the English language
Michael Unger,
graduate physician, consultant radiologist

David Stover
Publisher, Rock's Mills Press

James Hobson
Proofreading

Authors:

Prof. Dr. Siegfried Unger, graduate physicist, Berlin, Germany (June 17, 1925— October 19, 2015).

Jörn Schwarz, graduate engineer, member of the board, former Spokesman of the German Society of Refrigeration and Air Conditioning, DKV. Sponholz-Rühlow, Germany.

Michael Unger, graduate physician, consultant radiologist, transcriber and editor of the English text. Berlin, Germany.

Publisher of this edition:
David Stover, **Rock's Mills Press**. Oakville, Ontario, Canada.

Proofreading:
James Hobson, analytical chemist. London, United Kingdom; Berlin. Germany.

Edmund Altenkirch: Pionier der Kältetechnik
Original German edition published by
German Society of Refrigeration and Air Conditioning, DKV
Striehlstraße 11, DE 30159 Hannover, Germany
Tel.: +49 (511) 897 0814
Fax: +49 (511) 897 0815
E-mail: info@ DKV.org
www.DKV.org

ENGLISH EDITION PUBLISHED BY
Rock's Mills Press
Oakville, Ontario, Canada

Second Revised Edition, First English Edition

Table of Contents

1 Preface

Edmund Altenkirch was one of the great pioneers of refrigeration technology. In 1930 he was awarded, in appreciation of his outstanding contributions to the field of reversible heat and cold generation, especially in motionless and multistage absorption refrigeration machines, the academic degree of Doctor of Engineering "honoris causa[1]" by the Technische Hochschule Karlsruhe, Germany, and in the year 1950 he was awarded the highest accolade of the Deutscher Kälte- und Klimatechnischer Verein (DKV e.V.)[2], the Linde-Denkmünze[3]. Furthermore, for 12 years, he was editor-in-chief of the periodical "Zeitschrift für die gesamte Kälteindustrie," and, since 1936, had been the president of the seventh commission for research and education of the Institute International du Froid (Paris). His work and his significance in the field of refrigeration technology deserve a vigorous commemoration, which we hope the book will trigger.

The impulse to compile this work came from Reinhard Buchmann, who worked in the early 1950s at the III. Physikalisches Institut der Humboldt-Universität zu Berlin under the leadership of Prof. Eder. At this time, Siegfried Unger was a scientific research assistant there and concurrently Altenkirch's collaborator in his research establishment in Neuenhagen near Berlin. In the early 2000s Mr. Buchmann established contact between the DKV and Siegfried Unger and proposed honoring Edmund Altenkirch's diverse technical pioneering work in the form of a publication. The board of the DKV agreed with this proposal, and so the German edition of this treatise ensued, which appeared as one of a series of DKV status reports.

From 1946 until his death in the year 1953, Edmund Altenkirch had opened himself to his closest collaborator, Siegfried Unger, so that the latter could well empathize with Altenkirch's thinking and his motivations. Making use of his own publication[117], "E. Altenkirch: Grundgedanken und Ergebnisse seiner bedeutensten Arbeiten auf dem Gebiet der technischen Anwendungen der Thermodynamik"[4], Unger reviewed this work and, with the assistance of Jörn Schwarz, graduate engineer and member of the board of the DKV, the final treatise was written.

The authors owe much to the board as well as to the former executive secretary of the DKV, Irene Reichert, for their interest and support.

Prof. Dr. F. X. Eder, in those days Director of the Third Department of Physics at the Berlin Humboldt University, and the Department of Low Temperature Research at the former Academy of Science of the GDR[5] as well as later at the Walther-Meißner Institute in Munich, ought to take credit for the energetic pursuit of Altenkirch's suggestions and ideas.

---

[1] In the German language: Dr.-Ing. E. h. (Ehrenhalber) [translator's note].

[2] German Society of Refrigeration and Air Conditioning, [German abbreviation:] DKV; source: http://www.dkv.org/index.php?id=115), retrieved 2013-02-12 [translator's note].

[3] Linde Memorial Medal [translator's note].

[4] Siegfried Unger: "E. Altenkirch: Fundamental ideas and results of his most important work in the field of technical applications of the thermodynamics"[117] [translator's note].

[5] German Democratic Republic: Due to the status quo founded communist second German state on the territory of the postwar Germany (i. e., East Germany) [translator's note].

For a critical examination of the manuscript we thank Prof. Michael Arnemann, former Chairman of the Board of the German Society of Refrigeration and Air Conditioning, DKV, Prof. Helmut Lotz, Michael Unger, graduate physician, as well as Christa Radatz. Altenkirch's grandson Wolfgang Altenkirch, graduate mathematician, and his wife Dr. Christel Altenkirch, attorney-at-law, granted access to some valuable sources for this report[28].

SIEGFRIED UNGER AND JÖRN SCHWARZ

Dedication

Dear Christa-Maria,

I dedicate this book to you, who took part in our mutual lives, over decades in weal and woe.

In largest gratitude,

THINE SIEGFRIED

November 24, 2010

2 Biography

Edmund Altenkirch, son of elementary teacher Emil Altenkirch and his wife Augustine, née Kaethe, was born at Leibsch (Spreewald[6]) on August 11, 1880. He began attending the Gymnasium[7] in Luckau at Easter 1891, moving on to another Gymnasium,[7] "Zum Grauen Kloster" in Berlin in 1895. The latter school was extremely demanding, and, afraid he would not be able to meet those demands, Altenkirch tried to take his own life on May 27, 1897, when he threw himself in front of a train in the forest southwest of Potsdam, a town close to Berlin. He survived, but lost his left arm as a result of the tragic accident. Afterward, he attended the Friedrich Werder Gymnasium[7] in Berlin, which he left with the Abitur[8] in 1902; the schools inspector[9,] Michaelis, praised him in a special address.

After Altenkirch's father's death in 1892, his mother drew a widow's allowance of only 250 marks[10] a year. The Gymnasium[7] made his studies possible by supporting him with scholarships and the assignment of private lessons.

2.1 Studies at the Berlin University (as of 1903)

At the Berlin University (later Humboldt Universität zu Berlin), Altenkirch studied mathematics and physics. As he himself acknowledged[45], in his mathematical studies he owed "a lot to the docents HENSEL, SCHOTTKY, H.A. SCHWARTZ, and FROBENIUS. But physics had a stronger attraction for me. I was fascinated especially by WARBURG's experimental lectures, and MAX PLANCK's lectures and exercises in theoretical physics."

After his university studies his primary interest was in technical physics, an established field of science at that time. Here at first he studied ZEUNER's textbook "Lehrbuch der Technischen Thermodynamik" and the technical journal ZEITSCHRIFT FÜR DIE GESAMTE KÄLTEINDUSTRIE edited by H. LORENZ, which had become famous.

2.2 Efficiency of thermoelectric heating/cooling (1909 – 1912)

Early in his career, studying at the Berlin University, he was already fascinated by the thermocouple. 'Soon, I encountered a crucial gap,' he stated; 'there was no thermodynamically justified calculation for direct conversion of heat into electricity as it takes place in thermopiles, an easily imaginable means both to produce electricity from heat (thermopile) by the SEEBECK effect and, vice versa, thermal heat (heat pump) from electricity by the PELTIÉR effect[45].'

Working as an independent scholar, he wrote the papers ([1], [2]) which culminated in admirably clear fundamental statements regarding the technical efficiency of the considered methods. The technical and economic evaluation showed that the ther-

---

[6] A part of north-eastern Germany named for "forest at the river Spree," known for its special distinctive geomorphology (postglacial moraine landscape, sandy soil), biosphere (lowland, bog habitat), culture (most of the Sorbs, a Slavic minority, live there), and recreational activities (a navigable water canal system for punts) [translator's note].

[7] Academic high school [translator's note].

[8] German high-school diploma, precondition for university entrance [translator's note].

[9] In Germany a schools inspector is a teacher who is working for a local government and supervises schools in the corresponding administrative district [translator's note].

[10] The former German currency [translator's note].

mocouples available in those days produced too little thermoelectric power (usefully set out in the scale of "effective thermoelectric power" that Altenkirch himself introduced).

The response to Altenkirch's papers led to the so-called Peltier technology for special, mostly minimal-cooling applications (see, for example, Justi[74]), and parallel to that use its development has also helped to give rise to the technological innovations needed to produce devices capable of generating larger amounts of thermoelectric power.

2.3 Reversible heat increase (1913 – 1932)

After these initial investigations, Altenkirch focused his interest on analyzing the economic feasibility of various cooling and heating systems. To begin with, he tried to analyze the increase in heat production in systems driven by piston and lifting elements found in steam engines coupled with compression machines.

But the all-too-visible motions of the piston and lifting elements of the refrigeration machine, as well as the added complication introduced by the mutual compensation between generator and compressor, troubled him greatly. He sought therefore to achieve the desired heat transformation with simpler means, without visible mechanical motion and with a minimum of moving parts. His quest led him to rediscover Carré's absorption refrigeration device, which had first been invented 50 to 60 years earlier, but against whose acceptance there existed numerous prejudices.

Altenkirch proved the critics wrong with his Pioneer Patent D.R.P. 278 076[160], and the long paper "Reversible Absorptionsmaschinen"[2] in the periodical "Zeitschrift für die gesamte Kälteindustrie," in 1913 and 1914.

By simple structural measures, internal heat exchange, and the use of multiple stages he reached a new order of magnitude for the thermal ratio, one which had hitherto been considered impossible. A further generalization led Altenkirch eventually to incorporate a multistage design into the absorption machines which became universal heat transformers with thermal ratios up to 2 and beyond.

The use of thermal transformation for reversible heating applications that Altenkirch's research had shown to be feasible went on to be developed in complicated ways in the years that followed, which included the hyperinflationary period of the Weimar Republic. For instance, the application of heat pumps was further pursued only in Switzerland at the company Escher-Wyss. Altenkirch himself tackled the utilization of the temperature difference between outdoor cold and groundwater, or terrestrial heat for reversible heating—(heat pump)—with the West Berlin company (Dr. VOGELER) only after World War II.

2.4 The motionless absorption machine (1920 – 1953)

Altenkirch's innovative ideas for refrigeration became common knowledge during this period. Further research led to the enlargement of the thermal ratio and the realization of deep cooling applications (BORSIG BERLIN for grid gas cooling in the Ruhr district in Germany, MAIURI for the production of solid carbon dioxide ("dry ice") in England). Despite these successes, Altenkirch continued to improve the absorption machine, and set himself the task of eliminating the last moving parts—pumps and valves.

He called this new type of machine a "Kryotherme"[11], a word that made its way generally into the German technical vocabulary (Plank-Kuprianoff[97] 1948). In it, liquid columns replaced pumps and control valves. The patent for this device[137], followed that which covered the fluid cycle in absorption machines by means of gas bubble transportation[140]. In the latter case, because of the high pressure level in the vapor pressure curves of the cryogenic range of temperature, the two substances used—ammonia and water—required, in addition to the pressure maintenance by liquid columns, pressure equalization by carrier gas[12], according to GEPPERT[162].

In order to minimize the protraction of cold by the carrier gas cycle, Altenkirch lowered the pressure level by using the principle of resorption within the machine. This allowed him to reduce the required amount of carrier gas and to increase the proportion of the pressure maintenance that was achieved through the use of liquid columns. By reducing the internal pressure to normal atmospheric level, it was even possible to substitute for steel other materials, like glass, ceramics, or stonework.

An ideal field for application of these principles was the refrigerator, which had attracted rapidly growing worldwide interest. As a result, Altenkirch entered into a contract with the company SIEMENS-SCHUCKERT-WERKE (SSW). ELEKTROLUX[13] was pursuing parallel researches at the same time and the result was, as Altenkirch reported, 'an exceedingly tough competition of patents with legal actions in many countries.' SSW's unsuccessful lawsuit against ELEKTROLUX in the German Reich's Supreme Court was a considerable setback for both Altenkirch and SIEMENS and made continuation of work on the project impossible.

In the United States, however, competing companies failed to overturn the primacy of Altenkirch's patents, and instead paid nearly one million marks to SIEMENS for use of those patents. About 10 million absorption refrigerators using Altenkirch's system were produced.

2.5 Reversible air humidification and dehumidification (1932 – 1939)

By 1932 Altenkirch was searching for a new field of work, one which, in today's language, might be considered "green technology." This was the result of his early intention to use solar energy for heat transformation, a process that he achieved by using solid sorbents in an open circuit in combination with the simplest working media, namely, air as carrier gas and water as working medium.

The open-circuit process led him to another important generalization about heat transformation, which encompassed the chemical potential of water in the air. Altenkirch were granted a further pioneer patent[146] which dealt with the production of cold, of dry air, and of water extraction with the aid of solid sorbents and a patent[145] for a *simplified* method of water extraction, which he was able to test successfully in his research laboratory in Neuenhagen[123] near Berlin. Unfortunately, World War II put an abrupt end to these scientific investigations.

Section 8 is devoted to his research and development after World War II.

---

[11] In the text that follows, the English term "cryothermal apparatus" (Karl Stephan, "International Journal of Refrigeration," 1983) is used for Altenkirch's German term "Kryotherme" [translator's note].

[12] Also known as inert or auxiliary gas [translator's note].

[13] In 1957 the spelling of the company name was changed into Electrolux (source: http://de.wikipedia.org/wiki/Electrolux), retrieved 2013-02-09 [translator's note].

3 Thermocouple for Reversible Heating

Altenkirch was the first to consider the possibility of harnessing the thermoelectric effect to use lower-temperature sources (air, groundwater, geothermal energy) to provide heat. He even considered the idea of using the effect of electrothermal cooling and began experimenting with this early on, while still a *private scholar* with no formal qualifications.

As is generally known, reversible heat generation, as in the case of heat pumping, and refrigeration are nothing more than two distinct applications of one phenomenon. After the release of his highly praised paper, "Über den Nutzeffekt der Thermosäule"[1][14] Altenkirch concentrated his attention on the effectiveness of heat production in the paper "Elektrothermische Kälteerzeugung und reversible elektrische Heizung"[2][15]. In the latter publication he thanked his teacher H. Lorenz[78] who drew attention to the electrothermal refrigeration very early on, as it has a special position in engineering because it occurs without the involvement of a material body.

Later in this paper he placed emphasis on the fact that 'this peculiarity has the great benefit that a third medium is no longer required to generate cold or heat. In view of this benefit at times a higher wattage is justified. In addition, due to the electrothermal method the wattage for electric heating can be also be reduced, unlike the irreversible heating with Joule heat. It is therefore of both considerable theoretical and practical interest to demonstrate the relationship between the electrothermal refrigerating capacity and the expense of electric work as well as the investigation of the potential for saving electric energy for heating'.

3.1 Physical Effects

In 1821 Thomas Johann Seebeck (1770–1831) discovered the thermoelectric effect in an electric circuit of two conductors of different materials (legs) which were soldered together. With different temperatures at the solder junctions[16] (T_h = temperature at the hot thermojunction, T_o = temperature at the cold thermojunction), an electromotive force (*EMF*) originates in the circuit according to the relation:

$$EMF_{Seebeck} = \epsilon \cdot (T_h - T_o) \qquad \text{(Eq. 3.1)}$$

The proportionality factor ϵ [V·K⁻¹] is called thermoelectric power. In 1834 Jean Charles Athanase Peltiér (1785-1845) described the inverse effect (which was named for him) that in a heterogeneous circuit a flowing electric current I is accompanied by generation or absorption of a heat flow $\dot{Q}$ [J·s⁻¹] at the thermojunctions.

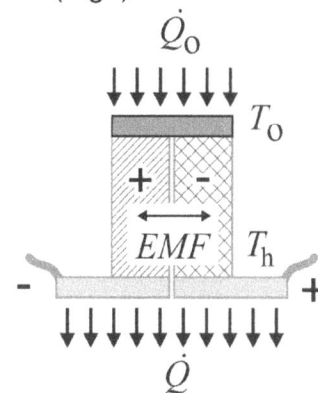

Fig. 3.1: Thermocouple according to Joffe

---

[14] "On the useful effect of the thermopile" [translator's note].

[15] "Electrothermal cold production and reversible electric heating" [translator's note].

[16] In the following text, the term "solder junction" is replaced by "thermojunction", cf. [175] [translator's note].

Due to the Peltiér effect, the thermocouple outlined in Figure 3.1 becomes a cooling aggregate which absorbs a heat flow (cold) at the cold thermojunction at temperature T_o and, after adding the supplied electric energy, emits it again as a warm heat flow at the hot thermojunction at a higher temperature T_h. The absolute value of the heat flow at temperature T_h follows the relation:

$$\dot{Q}_{\text{Peltiér}} = \Pi(T) \cdot I \quad \text{whereby the Equation 3.3 is valid.} \tag{Eq. 3.2}$$

The underlying thermodynamic interrelationships were formulated by WILLIAM THOMSON (1824-1907)[17]. According to Thomson a relation exists between the coefficients Π and $\mathcal{E}$ of the Peltiér and Seebeck effect shown in Equation 3.2:

$$\Pi = T \cdot \mathcal{E}_{\text{abs}} \tag{Eq. 3.3}$$

where the Peltiér heat flow[18] is described by

$$\dot{Q}_{\text{Peltiér}} = \Pi \cdot I \tag{Eq. 3.4}$$

3.2   Thermodynamic Interpretation[19]

By contrast with Altenkirch's predecessors, J. KOLLERT[76] and H. HOFFMANN[70], he immediately developed a thermodynamic interpretation of the thermoelectric phenomena, which was based on the First and Second Laws[20]. In summary his line of thought was as follows. Let:

- P be the reversible wattage the current I transfers electrically,
- $\Pi_h \cdot I$ be the emitted Peltiér heat (Π_h = Peltiér coefficient) at the hot thermojunction at temperature T_h, and
- $\Pi_o \cdot I$ be the absorbed heat (Π_o = Peltiér coefficient) at the cold thermojunction at temperature T_o,

so, with the reversible process conduct, the First and Second Law are valid in the form

Fist Law: $$W = \Pi_h \cdot I - \Pi_o \cdot I = \mathcal{E} \cdot (T_h - T_o) \cdot I \tag{Eq. 3.5}$$

Second Law: $$\frac{\Pi_h \cdot I}{T_h} = \frac{\Pi_o \cdot I}{T_o} \quad \Leftrightarrow \quad \frac{\Pi_h}{T_h} = \frac{\Pi_o}{T_o} \tag{Eq. 3.6}$$

After resolution to Π_o and Π_h these two equations can be written

$$\Pi_o = \Pi_h - \mathcal{E} \cdot (T_h - T_o) \quad \text{and} \quad \Pi_h = T_h \cdot \frac{\Pi_o}{T_o} \tag{Eqs. 3.7}$$

---

[17] Knighted in 1866, elevated to the peerage as "Lord Kelvin", 1892.

[18] The absolute thermoelectric power of the leg materials, which occurs in Equation 3.3 (after integration of Equation 3.3), can also be determined as a measurement against whichever etalon (e. g. platinum). Meanwhile, the theoretical relations in Equations 3.3 to 3.6 have been substantiated by the "Thermodynamics of Irreversible Processes" (especially by the "Onsager Relations")[115], [65].

[19] Readers who would prefer to skip the mathematical derivations here may proceed to the Summery of Results at the end of the section.

[20] Nowadays in a "thermostatic" indicated description (see DeGroot, S. R.[65]).

Through insertion of Π_h into Π_o the result is:

$$\Pi_o \cdot \frac{T_h - T_o}{T_o} = \mathcal{C} \cdot (T_h - T_o) \qquad \text{(Eq. 3.8)}$$

After rewriting the equation, the *Peltiér* relations follow immediately:

$$\Pi_o = \mathcal{C} \cdot T_o \quad \text{or generally} \rightarrow \quad \Pi = \mathcal{C} \cdot T \qquad \text{(Eqs. 3.9, 3.10)}$$

3.3 Altenkirch's Thermophysical Approach

3.3.1 Internal Irreversibilities

Heat conduction in the legs from the hot to the cold thermojunction
The legs are an (undesired) heat bridge between the hot and the cold thermojunction where their heat conducting values $L_{\lambda a} = A \cdot \lambda_a$ and $L_{\lambda b} = B \cdot \lambda_b$ formed by the geometry factors A, B (quotient of cross-section and length of the legs a, b) and their specific thermal conductivities λ_a and λ_b is a thermal parallel connection. The heat flow which flows over the bridge and the heat conductance total L_λ is thus given by

$$\dot{Q}_\lambda = L_\lambda \cdot (T_h - T_o) \quad \text{with} \quad L_\lambda = L_{\lambda a} + L_{\lambda b}, \quad L_{\lambda a} = A \cdot \lambda_a, \quad L_{\lambda b} = B \cdot \lambda_b \qquad \text{(Eqs. 3.11)}$$

Joule heat generation in the legs
This effect, which is also undesired, is proportional, respectively, to the electric resistance total $[\Omega = A^{-1} \cdot V]$ of the thermocouple, and therefore, to the electrical serial connection of the *legs' resistances*.

Starting from the specific thermal conductivities σ_a and σ_b and the electric conducting values $L_{\sigma a}$, $L_{\sigma b}$ with the dimension of the legs a, b formed analogously according to the heat conducting values, so the electric resistance R of the serial connection of the legs and the Joule heat generation total is obtained as

$$\dot{Q}_\sigma = I^2 \cdot R, \quad R = L_{\sigma a}^{-1} + L_{\sigma b}^{-1} \quad \text{with} \quad L_{\sigma a} = A \cdot \sigma_a \quad \text{and} \quad L_{\sigma b} = B \cdot \sigma_b \qquad \text{(Eqs. 3.12)}$$

3.3.2 External Effects[21]

In order to determine the heat flows $\dot{Q}_o$ and $\dot{Q}_h$, exchanged with the external media, the thermal balances at the thermojunctions need to be set up. To the environment the balance $\dot{Q}_h = \dot{Q}_o + P$ is valid. Therein P is the supplied wattage total and $\dot{Q}_o$, $\dot{Q}_h$ the heat flows from and to the thermojunctions with the temperatures T_o and T_h. The heat flow $\dot{Q}_\lambda$ by heat conduction removes the same share of the absolute value of both the refrigerating capacity at T_o and the thermal output at T_h.

Considering the Joule heat flow $\dot{Q}_\sigma$, Altenkirch demonstrated that it is distributed on both thermojunctions in equal shares and thus it enlarges the heat flow at T_h but lowers the refrigerating capacity at T_o. Considering the *Peltiér* heat flows according to Equation 3.10 the thermal balance equations at the thermojunctions are

---

[21] DOUGLAS and JOFFE [61], [71] deviate from Altenkirch's justified optimal relation $\alpha = \alpha_o$ ($\alpha = A/B$; A, B: cross-section/length of the legs a, b, see Nomenclature) as it may perhaps be required to attain the balance between economical feasibility and technical optimality.

$$\dot{Q}_o = \Pi_o - \dot{Q}_\lambda - \frac{\dot{Q}_\sigma}{2}, \quad \dot{Q} = \Pi_h - \dot{Q}_\lambda + \frac{\dot{Q}_\sigma}{2} \quad \text{and} \quad P = \dot{Q}_h - \dot{Q}_o \qquad \text{(Eqs. 3.13, 3.14, 3.15)}$$

The net Peltiér heat flows at the thermojunctions according to Equations 3.13 to 3.15 depend, apart from their specific material properties λ, σ, ϵ and the temperatures T_h, T_o only on the electric current I and the geometry factor A and B whereas the external influences of the thermocouple as a cold or heat performing unit are determined, above all, by the *cross-sections of the legs* at the thermojunctions and the mode of transmission of heat to external media like cooling chambers, ice bunkers, and cooling water. These processes underlie the Laws of Heat Transfer, for example, to liquefied media and/or the Law of Thermal Conduction via metallic bridge elements to/from finned conductors of heat, and the like.

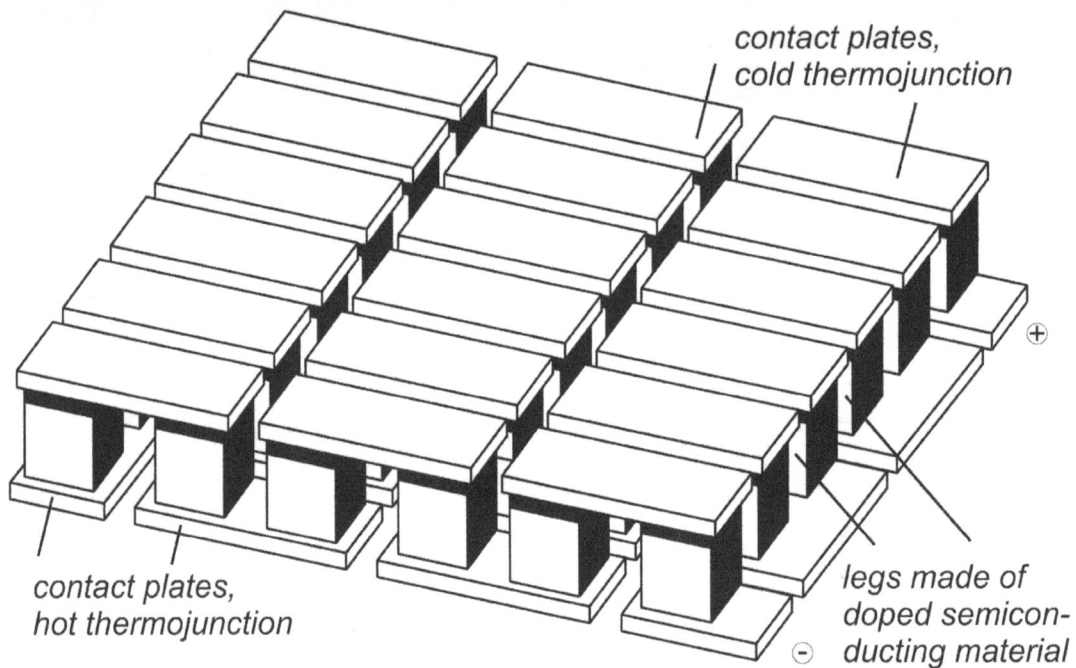

Fig. 3.2: Thermal battery: Scheme of a block of thermocouples connected electrically in series; by Cube, L. v. (1997) (editor), paper by H. Lotz[56]

An essential practical external condition for electric consumers is, of course, the limitation of electrical connections to common dimensions in order to avoid unfavorably large cable cross-sections for electric input leads (and associated with it large electric fuse protections). As a result, flat dispositions of thermocouples (arrays, batteries) are favored in which the single elements are connected electrically in series but thermally in parallel (see Figure 3.2).

3.4   Threshold Values for Efficiency and Expenditure

Aside from the specific material properties λ, σ, the thermoelectric power ϵ and the temperatures T_h, T_o at the thermojunctions, the external influences of the thermocouple furthermore depend only on the electric current I and the geometry factors A

and B of the legs. With equal values of the temperatures T_h, T_o, equal substances, and equal geometry the amount of the electric current I remains variable. The following dependencies exist:

- the heat flows $\Pi_h \cdot I$, $\Pi_o \cdot I$ (warm and cold) will be both enhanced proportionally with I
- but the cold loss $\dot{Q}_o$ grows with the square of I,
- the cold/heat loss $\dot{Q}_\lambda$ developed by heat conduction remains however constant.

Therefore, there is a wattage in which the cold $\dot{Q}_o$ reaches a maximum. On this basis Altenkirch provided a complete theory of thermoelectric refrigeration and reversible electrothermal heating. Especially important theoretical issues for applications in refrigeration technology taken up and treated thoroughly by him are:

- the maximum of refrigeration and the lowest reachable temperature of the refrigerating capacity (maximum temperature lift)
- the minimum expense for refrigeration → maximum Coefficient of Performance (COP)
- the enlargement of the temperature lift by cascade design.

Next follows a description for the refrigeration; an example for warming is not given since refrigeration and thermal heat production are only two different aspects of the same phenomenon.

3.4.1 Lowest Reachable Temperature of Refrigeration

The detailed formula for the refrigerating capacity (Equations 3.13 to 3.15) by taking into account the loss effects in Equations 3.11 and 3.12 and the expression for Π_o according to Equation 3.9 is

$$\dot{Q}_o = \epsilon \cdot I \cdot T_o - L_\lambda (T_h - T_o) - \frac{1}{2} \cdot R \cdot I^2 \qquad \text{(Eq. 3.16)}$$

The lowest temperature $min\ T_o$ of the refrigerating capacity is reached with increase (or alteration) of the electric current up to the I maximum of $\dot{Q}_o$ when the maximum just disappears, that means, at $max_1 \dot{Q}_o = 0$. Thereby, the temperature T_h at the hot thermojunction has to be kept constant. The enlargement of the electric current causes, on one hand, an enhancement of the net cold by the Peltiér cold, and, on the other hand, a decrease by the Joule heat. An equilibrium between both is reached according to

$$I_{max\ q_o} = \frac{\epsilon \cdot T_o}{R} \qquad \text{(Eq. 3.17)}$$

Through insertion, the expression for the maximum cold is obtained

$$max_1 \dot{Q}_o = \frac{1}{2} \cdot \frac{\epsilon^2 T_o^2}{R} - L_\lambda (T_h - T_o) \qquad \text{(Eq. 3.18)}$$

The reachable lower threshold value for T_o has yielded after resolution of the equation for $max_1 \dot{Q}_o = 0$ toward T_o.

The solution is denoted with Θ_o. The intermediate step is shown by

$$\frac{T_h - \Theta_o}{\Theta_o^2} = \frac{\epsilon^2}{2 \cdot R \cdot L_\lambda}$$ (Eq. 3.19)

The resolution to Θ_o is not quite trivial as, again, the occurring product $R{\cdot}L_\lambda$ itself depends on the temperature.

According to Altenkirch's following contemplations for the approximation of this temperature dependence it was considered to simplify, instead of the dependence of the material constants on the temperatures T_o, T_h, only the dependence on the arithmetic mean tempertestature T_m at the thermojunctions, which in the following period had been adopted by most of the successive authors. According to that, the dependence of the material constants on the arithmetic mean temperature T_m of the thermojunctions is assumed as

$$T_m = \frac{T_h + \Theta_o}{2}$$ (Eq. 3.20)

The product is explicitly shown by

$$R \cdot L_\lambda = (A \cdot \lambda_a + B \cdot \lambda_b) \cdot \left(\frac{1}{A \cdot \sigma_a} + \frac{1}{B \cdot \sigma_b} \right) = \frac{\lambda_a}{\sigma_a} + \alpha \cdot \frac{\lambda_a}{\sigma_b} + \alpha^{-1} \cdot \frac{\lambda_b}{\sigma_a} + \frac{\lambda_b}{\sigma_b}$$ (Eq. 3.21)

The factor $\alpha{=}A/B$ can be optimized for the maximum cold. One obtains

(Eq. 3.22)

$$\alpha[max(\dot{Q}_o)] = \alpha_o = \sqrt{\frac{\sigma_b \cdot \lambda_a}{\sigma_a \cdot \lambda_b}}$$

Through insertion into Equation 3.21 it comes to

$$R \cdot L_\lambda = \left(\frac{\lambda_a}{\sigma_a} + \frac{\lambda_b}{\sigma_b} \right)^2 \text{ test}$$ (Eq. 3.23)

Herein Altenkirch approximated the quotients λ/σ, following the Wiedemann-Franz-Lorenz law

$$\left(\frac{\lambda}{\sigma} \right)_{id} = WFL = 2.5 \cdot 10^{-8}$$ (Eq. 3.24)

for ideal metals, with the WFL number—named for their authors[22]. He took into account the deviation from the ideal case (above mentioned) by a yet material-dependent, but in the range of temperature considered presupposed constant factor ρ in the form

---

[22] Gustav Wiedemann, Rudolph Franz, and Ludwig Lorenz [translator's note].

$$\rho = \frac{\dfrac{\lambda}{\sigma}}{\left(\dfrac{\lambda}{\sigma}\right)_{id}} \quad \rightarrow \quad \left(\frac{\lambda}{\sigma}\right) = \rho \cdot WFL \cdot T_m \qquad \text{(Eq. 3.25)}$$

so that therefore both legs' materials a, b are distinguished still only by their different "derivatives" ρ_a, ρ_b from the Wiedemann-Franz-Lorenz law. Through insertion of Equation 3.23 becomes according to Equation 3.25

$$R \cdot L_\lambda = WFL \cdot \frac{T_h + \Theta_o}{2} \cdot \left(\sqrt{\rho_a} + \sqrt{\rho_b}\right)^2 \qquad \text{(Eq. 3.26)}$$

Through insertion into Equation 3.19 after the resolution to Θ one yields

$$\frac{T_h - \Theta_o}{\Theta_o^2} = \frac{\mathcal{E}^2}{WFL \cdot (T_h + \Theta_o) \cdot \left(\sqrt{\rho_a} + \sqrt{\rho_b}\right)^2} \qquad \text{(Eq. 3.27)}$$

Then, from $max_1 \, \dot{Q} = 0$ according to Equation 3.18 it follows

$$\frac{T_h^2 - \Theta_o^2}{\Theta_o^2} = 0.25 \cdot WFL^{-1} \cdot \mathcal{E}'^2 \qquad \text{(Eq. 3.28)}$$

where $\mathcal{E}'$ denotes Altenkirch's so defined and named "effective thermoelectric power"

$$\mathcal{E}' = \frac{2 \cdot \mathcal{E}}{\sqrt{\rho_a} + \sqrt{\rho_b}} \cdot \qquad \text{(Eq. 3.29)}$$

$$\Psi = \sqrt{\left(1 + 0.25 \cdot WFL^{-1} \cdot \mathcal{E}'^2\right)} = \sqrt{\left(1 + 10^7 \cdot \mathcal{E}'^2\right)} \quad \text{results in} \qquad \text{(Eq. 3.30)}$$

$$\frac{T_h}{\Theta_o} = \Psi = \frac{\Theta}{T_o} \qquad \text{(Eq. 3.31)}$$

with Θ as a solution of the minimizing problem toward T, the lowest reachable temperature at the cold thermojunction, provided that the temperature of the hot thermojunction $= T_h$ is fixed.

Through the right side of Equation 3.31 the maximum reachable temperature Θ at the hot thermojunction is also determined. In the first case, when the cold thermojunction is fixed at temperature T, it is valid

$$\frac{T_h - \Theta}{T_h} = \frac{\Psi - 1}{\Psi} = \left(1 - \Psi^{-1}\right) \cdot T_h \quad \text{and} \quad \frac{\Theta - T}{T} = \Psi - 1 \qquad \text{(Eqs. 3.32, 3.33)}$$

The interrelationships are outlined graphically in Figure 3.3. The solid-line curve shows the temperature drop $T_h - \Theta_o$ below T_h starting from the temperature at the hot thermojunction $T_h = 283$ in accordance with Equation 3.32. The dotted curve describes the temperature increase $\Theta - T$ starting from T up to the temperature Θ according to Equation 3.33 for $T_o = 273$ K; both curves are nearly identical for small-sized effective thermoelectric powers.

The formula expression for the value of the dotted curve (in case of the heat pump: temperature lift from T_o up to the level Θ) has still specified

$$\Delta T = \Theta - T_o = (\Psi - 1) \cdot T_o = ((1 + 10^7 \mathcal{E}'^2)^{1/2} - 1) \cdot T_o \qquad \text{(Eq. 3.34)}$$

The formula expression for the y-coordinates of the solid-line curve (refrigeration as drop of the temperature from T_h to Θ_o) is shown by

$$\Delta T = T_h - \Theta_o = (1 - \Psi^{-1}) T_h = (1 - (1 + 10^7 \mathcal{E}'^2)^{-1/2}) T_h \qquad \text{(Eq. 3.35)}$$

According to the preceding the refrigeration is only enabled if the following nesting for the temperatures is given by

$$\Theta_o \leq T_o \leq T_h \leq \Theta \qquad \text{(Eq. 3.36)}$$

Voigt, H.[120] specified the following generalization of Equation 3.36 for firmly predetermined values $\dot{Q}_o, \dot{Q}_h$ of the required refrigerating capacities and heat outputs

$$\frac{T_h}{T_o} = \frac{\Psi \cdot \dot{Q}_h + \dot{Q}_o}{\Psi \cdot \dot{Q}_o + \dot{Q}_h} \qquad \text{(Eq. 3.37)}$$

With the drop of the refrigeration capacity toward zero, as it underlies Equation 3.32, Equation 3.37 can be simplified into Equation 3.32—in the Table (see Tab. 3.1: Numerical Example) a numerical example is calculated for an alloy pair of a thermocouple according to Plank[102] with the leg a: 38 Sb/62 Te and leg b: 63 Pb/37 Te. The substance values have obtained at $T_m = 253$ K, and the WFL has entered with $2.5 \cdot 10^{-8}$ [V^2·K^2].

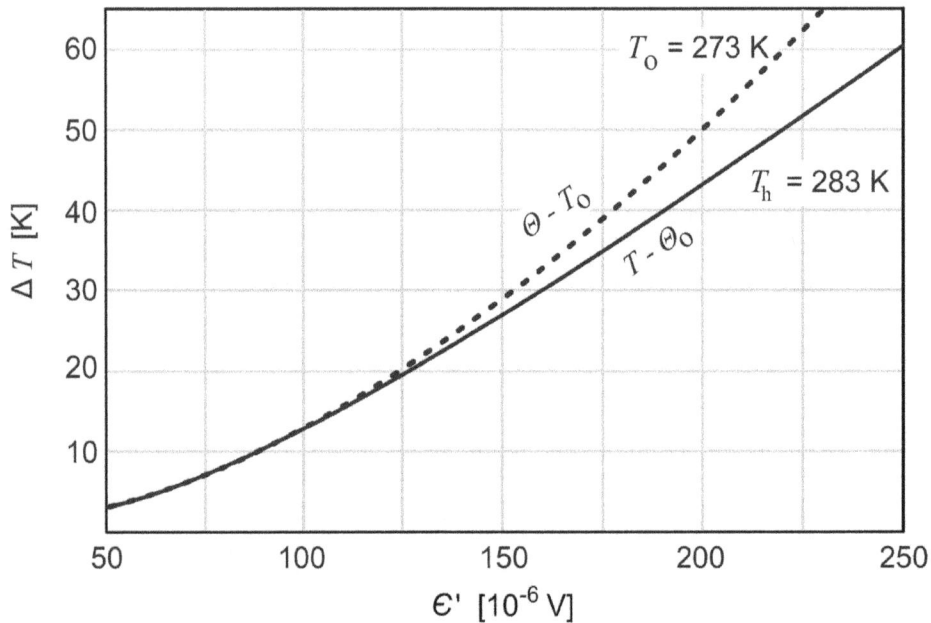

Fig. 3.3: Maximum temperature difference as a function of the effective thermoelectric power

Tab. 3.1: Numerical Example

Name	Symbol	Formula	Eq.	Value	Dim.
specific electrical conductivity for a	σ_a			$4.00 \cdot 10^5$	$[m^{-1} \cdot \Omega^{-1}]$
specific electrical conductivity for b	σ_b			$1.15 \cdot 10^5$	$[m^{-1} \cdot \Omega^{-1}]$
heat conductivity for a	λ_a			3.9	$[W/(m \cdot K)]$
heat conductivity for b	λ_b			4.0	$[W/(m \cdot K)]$
abs. thermoelectric power against Cu for a	$\epsilon_{abs\,[a]}$			$71 \cdot 10^{-6}$	$[V \cdot K^{-1}]$
abs. thermoelectric power against Cu for b	$\epsilon_{abs\,[b]}$			$-174 \cdot 10^{-6}$	$[V \cdot K^{-1}]$
thermoelectric power	ϵ	$\epsilon_{abs\,[a]} - \epsilon_{abs\,[b]}$		$0.245 \cdot 10^{-3}$	$[V \cdot K^{-1}]$
temperature of the hot thermojunction	T_h			293 K with refrigeration	$[K]$
temperature of the cold thermojunction	T_o			273 K with heat pump	$[K]$
Wiedemann-Franz-Lorenz number	WFL			$2.5 \cdot 10^{-8}$	$[V^2 \cdot K^{-2}]$
deviation from the WFL Law; A, B (at T = 283 °C)	$\rho_{a,b}$	$\dfrac{\left(\frac{\lambda}{\sigma}\right)_{a,b}}{WFL \cdot T}$	3.29	A: 1.37 B: 4.89	$[-]$
effective thermoelectric power [Altenkirch]	ϵ'	$\dfrac{2 \cdot \epsilon}{(\sqrt{\rho_a} + \sqrt{\rho_b})}$	3.33	$0.145 \cdot 10^{-3}$	$[V \cdot K^{-1}]$
efficiency [Altenkirch]	Ψ	$(1 + 10^7 \cdot \epsilon'^2)^{0.5}$	3.33	1.1	$[-]$
maximum temperature decrease	$(T-T_o)_{max}$	$\dfrac{T \cdot (\Psi - 1)}{\Psi}$	3.36	26.6	$[K]$
maximum temperature increase	$(T-T_o)_{max}$	$T \cdot (\Psi - 1)$	3.37	27.3	$[K]$
figure of merit [Lotz]	Z	$(R \cdot L_\lambda)^{-1} \cdot \epsilon^2$	3.42	$2.39 \cdot 10^{-5}$	$[K^{-1}]$

Neglect of the Temperature Dependence of L_λ / L_σ

At this section a note follows regarding H. Lotz's[56] mathematical treatment who introduces a simplification of Equation 3.18 through evaluation of Equation 3.28 by ignoring the explicit temperature dependence of the quotient on T_o:

$$max(T_h - T_o) = \frac{1}{2} \cdot Z \cdot T_o^2 \qquad \text{(Eq. 3.38)}$$

He called the following expression a quality criterion ("figure of merit")

$$Z = (R \cdot L_\lambda)^{-1} \cdot \epsilon^2 = \frac{L_\sigma}{L_\lambda} \cdot \epsilon^2 \quad [K^{-1}]$$ (Eq. 3.39)

3.4.2 Maximum Coefficient of Performance (COP) of Cooling

Another issue was the question about the efficiency of refrigeration. In consideration of this Altenkirch took as a basis again the optimal ratio α_o of the geometry factors A, B. With $\dot{Q}_o$ according to the Equations 3.11 and 3.16

$$\dot{Q}_o = \epsilon \cdot I \cdot T_o - \left(\dot{Q}_\lambda = L_\lambda \cdot (T_h - T_o)\right) - \frac{1}{2} \cdot R \cdot I^2 \quad \text{and}$$ (Eq. 3.40)

$$\dot{Q}_h = \epsilon \cdot I \cdot T_h - \left(\dot{Q}_\lambda = L_\lambda \cdot (T_h - T_o)\right) + \frac{1}{2} R \cdot I^2$$ (Eq. 3.41)

becomes

$$P = \dot{Q}_h - \dot{Q}_o = \epsilon \cdot I (T_h - T_o) + R \cdot I^2$$ (Eq. 3.42)

and thus the expression for the COP of cooling

$$\epsilon_o = \frac{\dot{Q}_o}{P} = \frac{\epsilon \cdot I \cdot T_o - L_\lambda (T_h - T_o) - \frac{1}{2} R \cdot I^2}{\epsilon \cdot I \cdot (T_h - T_o) + R \cdot I^2}$$ (Eq. 3.43)

The maximization according to I leads to

$$\epsilon_o = \epsilon_{o\,max\,I} = \epsilon_{o\,id} \frac{\Psi + 1}{\Psi - \dfrac{T_h}{T_o}}$$ (Eq. 3.44)

with the CARNOT factors (the theoretical maximum COP's) of refrigeration and generation of thermal heat, respectively:

$$\epsilon_{o\,id} = \frac{T_o}{T_h - T_o} \quad \text{and} \quad \epsilon_{h\,id} = \frac{T_h}{T_h - T_o}$$ (Eq. 3.45)

thus, the maximum quality grade of refrigeration is shown by

$$\eta_o = \frac{\epsilon_{o\,max\,I}}{\epsilon_{o\,id}} = \frac{\psi - \tau}{\psi + 1} \quad \text{with} \quad \tau = \frac{T_h}{T_o} \text{[23]}$$ (Eq. 3.46)

and, generally is shown by

$$quality\ grade \quad \eta = \frac{real\ COP\ \ \epsilon}{ideal\ COP\ \ \epsilon_{Carnot}} \leq 1$$ (Eq. 3.47)

$\eta_o = 1$ corresponds to the ideal case, otherwise it is $0 \leq \eta \leq \eta_o \leq 1$.

---

[23] in the following text of this section $\eta_{o\,max\,I}$ is always written as η_o

Ψ is growing with $\mathcal{E}'$. The increase of Ψ has a greater impact on the numerator than on the denominator of Equation 3.46, thus, the quality grade η_o enlarges with a growing Ψ. For $\mathcal{E}' = \infty$ is $\Psi = \infty$, thus $\eta_o = 1$. In the latter case the quality grade is directly given by the (ideal) CARNOT factor.

With little thermoelectric powers, that means, when Ψ goes toward 1, η_o aims toward 0; this, because of the disappearing Peltiér cold with maintained constant losses through Joule heat and heat conduction; the dotted curves in Figure 3.4 are defined for infinite cascade connections (see the next paragraph but one).

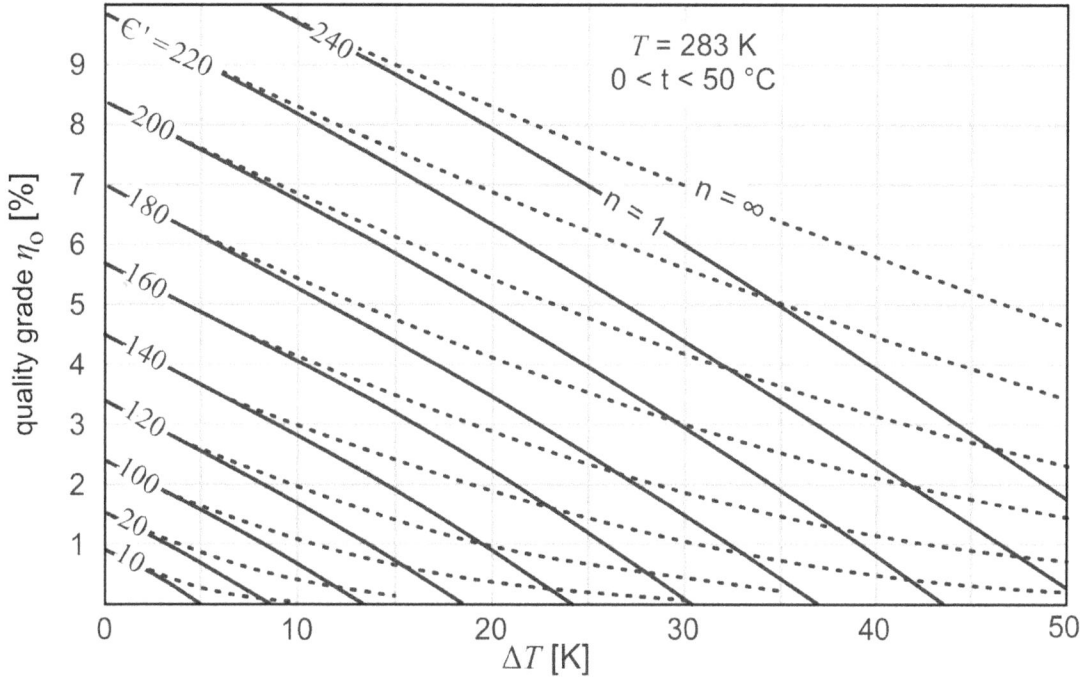

Fig. 3.4: Quality grade of the thermoelectric refrigeration

3.4.3 Processes at the hot Thermojunction

According to the First Law it is defined

$$Q_h = P + Q_o \qquad\qquad (Eq.\ 3.48)$$

With the wattage P the COP of heating results in

$$\mathcal{E}_h = \frac{\dot{Q}_h}{P} = \frac{P + \dot{Q}_o}{P} = 1 + \mathcal{E}_o \qquad\qquad (Eq.\ 3.49)$$

There, the maximum quality grade of the heat generation follows analogously to η_o

$$\eta_h = \mathcal{E}_h \cdot \frac{T_h - T_o}{T_h} \qquad\qquad (Eq.\ 3.50)$$

The conditions of the smallest expenditure of work for the heat generation are exactly the same as for the refrigeration. Figure 3.3 displays the temperature differences up to which the saving on energy is still easily enabled. According to the preceding, the relative saving on energy with heating in contrast to heating with Joule heat is shown by

$$\frac{\dot{Q}_h - P}{\dot{Q}_h} = \frac{\epsilon_h}{1 + \epsilon_h}$$

(Eq. 3.51)

When η_o ($0 \le \eta_o \le 1$) is large, the saving on energy can be very significant, especially for raising the heat to slightly higher temperatures. But a saving always takes place as long as the cold thermojunction is still capable to extract heat from the environment. As a limit for that, the temperature difference: $T_h - \Theta_o$ has been found (see Figure 3.3).

3.4.4 Cascade Design

Altenkirch showed that this threshold can be moved out *arbitrarily far* by cascade design. The thermodynamics of cascade design for heat pumps was later (in 1920) covered by Altenkirch in a compact and comprehensive form[12] which we would like to refer to below.

For the mathematical treatment Altenkirch introduced there the term *quality grade of the thermal ratio* on a heat pump which takes up a thermal power $\dot{Q}_o$ at T_o and an electric power (wattage) P and emits a heat output $\dot{Q}_h$ at T_h.

According to Equation 3.51, $\dot{Q}_o / \dot{Q}_h$ was the achieved saving on reversible compared to irreversible heating. For, with $\dot{Q}_o = 0$ the thermal heat taken from the heat reservoir at T_o would be = 0, thus, the entire released heat output $\dot{Q}_h$ at the higher temperature $T = T_h$ had only to be brought up by the (irreversible) electric resistance heating, that means, through the annihilation of wattage

$$P = \dot{Q}_h - \dot{Q}_o$$

(Eq. 3.52)

The maximum: $\dot{Q}_o / \dot{Q}_h = 1$ is thermodynamically impossible; the maximum limit value therefore is, as it is known

$$\left(\frac{\dot{Q}_o}{\dot{Q}_h}\right)_{id} = \tau^{-1} \quad , \quad \tau = \frac{T_h}{T_o}$$

(Eq. 3.53)

Therefore Altenkirch describes this saving in the form

$$\left(\frac{\dot{Q}_o}{\dot{Q}_h}\right) = \eta' \cdot \tau^{-1}$$

(Eq. 3.54)

and calls η' the *"quality grade of the thermal ratio"*. There, η' measures the *degree of the approximation to the ideal value T_o / T_h of the thermal ratio $\dot{Q}_o / \dot{Q}_h$. It is given by: $\eta' = 0$ (irreversible), $\eta' = 1$ (reversible)*—η' can be reduced to the quality grade η_o of refrigeration. The mathematical derivation, in brief, is shown by:

Starting from η_o as the corresponding ratio of the real COP: $\epsilon_o = \dot{Q}_o / P$ to the ideal one (Equation 3.46), it is yielded by application of the thermal balance Equation $\dot{Q}_h = \dot{Q}_o + P$ in consideration of Equation 3.53 due to elemental transformations

$$\frac{\dot{Q}_o}{\dot{Q}_h} = \frac{\dot{Q}_o / P}{1 + \dot{Q}_o / P} = \frac{\epsilon_o}{1 + \epsilon_o} = \frac{\eta_o \epsilon_{id\,o}}{1 + \eta_o \epsilon_{id\,o}} = \frac{\eta_o T_o}{T_h - T_o + \eta_o T_o} \qquad \text{(Eq. 3.55)}$$

and furthermore

$$\frac{\dot{Q}_o}{\dot{Q}_h} = \frac{\eta_o}{\tau \cdot (1 - \eta_o)} \qquad \text{(Eq. 3.56)}$$

from which by comparison with Equation 3.54 the following is derived

$$\eta' = \frac{\eta_o}{1 - \tau^{-1} \cdot (1 - \eta_o)} \qquad \text{(Eq. 3.57)}$$

For the quality grade of the thermocouple the following formula is inferred from Equation 3.46

$$\eta_o = \frac{\Psi - \tau}{\Psi + 1} \qquad \text{(Eq. 3.58)}$$

Therewith, it is obtained $\quad \eta' = \dfrac{\Psi - \tau}{\Psi - \tau^{-1}} \qquad \text{(Eq. 3.59)}$

The heat outputs, refrigerating capacities, and temperatures appearing on the cascade out of n elements $E_1, \ldots, E_n$ be as follows

Begin	Element 1	Element 2		Element n	End
$\dot{Q}_o = \dot{Q}_{10}$	$\dot{Q}_{10} = \dot{Q}_{1h} \rightarrow$	$\dot{Q}_{20} = \dot{Q}_{2h} \rightarrow$	$\cdots$	$\dot{Q}_{n0} = \dot{Q}_{nh} \rightarrow$	$\dot{Q}_h = \dot{Q}_n$
$T_o = T_{10}$	$T_{10} = T_{1h}$	$T_{20} = T_{2h}$	$\cdots$	$T_{n0} = T_{nh}$	$T_h = T_n$

By transferring ($\rightarrow$) a heat flow from the warm side of an element to the cold side of the following one a cascade develops. The first element should produce the refrigerating capacity $\dot{Q}_o$ and the nth the wattage $\dot{Q}_h$.

Altenkirch pointed out that according to Equation 3.57 η' lies much closer to 1 than the single quality grades of refrigeration and heat production η_o and η_h, that means, if the temperature difference being 30 °C as is the case with $T_o = 273\ °K$ and $T_h = 303\ K$ ($\tau_h^{-1} = 0.9$), the value of η' for $\eta_o = 0.7$ results to

$$\eta' = \frac{0.7}{1 - 0.9 \cdot (1 - 0.7)} = 0.96 \qquad \text{(Eq. 3.60)}$$

For the balance at the cascade the "quality grades of the thermal ratios" at the single stages are introduced by η_i', $i = 1 \ldots n$. At first, due to the heat transfers from stage to stage follows

$$\frac{\dot{Q}_o}{\dot{Q}_h} = \frac{\dot{Q}_o}{\dot{Q}_1} \cdot \frac{\dot{Q}_1}{\dot{Q}_2} \cdot \frac{\dot{Q}_2}{\dot{Q}_3} \cdots \frac{\dot{Q}_{n-1}}{\dot{Q}_n} \cdot \frac{\dot{Q}_n}{\dot{Q}_h}$$ (Eq. 3.61)

Here, according to Equation 3.54 it can be substituted for the single stages

$$\frac{\dot{Q}_o}{\dot{Q}_h} = \eta_1' \cdot \frac{T_o}{T_1} \cdot \eta_2' \cdot \frac{T_1}{T_2} \cdot \eta_3' \cdot \frac{T_2}{T_3} \cdots \eta_{n-1}' \cdot \frac{T_{n-1}}{T_n} \eta_n' \cdot \frac{T_n}{T_h}$$ (Eq. 3.62)

After reducing the internal temperature quotients follows

$$\frac{\dot{Q}_o}{\dot{Q}_h} = \tau_{whole}^{-1} \cdot \left(\Pi_{i=1\ldots n} \, \eta_i \right) \text{ with}$$ (Eq. 3.63)

$$\tau_{whole} = \Pi_{i=1\ldots n} \tau_i \text{ at } \tau_i = \frac{T_i}{T_{i-1}} \text{ and } T_h = T_{i=n} \, , \, T_o = T_{i=0}$$ (Eq. 3.64)

and thus for the "quality grade of the thermal ratio at the whole cascade" the product

$$\eta' = \Pi_{i=1\ldots n} \, \eta_i'$$ (Eq. 3.65)

Analogously, the temperature ratios are shown by

$$\tau_{whole} = \Pi_{i=1\ldots n} \tau_i \text{ with } T_h = T_{h(i=n)} \text{ and } T_o = T_{o(i=1)} \, , \, \tau_i = \frac{T_{hi}}{T_{oi}}$$ (Eq. 3.66)

If one selects the same-sized temperature quotients T_{hi}/T_{oi} at the single stages of the cascade ($= \tau$), so from Equation 3.63 follows

$$\tau_i = \left(\tau_{whole} \right)^{1/n} \text{ for all } i$$ (Eq. 3.67)

Herein, τ_{whole} is the whole temperature quotient (lift) which has to be bridged by the cascade, and it comes to

$$\eta_i' = \frac{\eta_o}{1 - \tau_{whole}^{-1} \cdot (1 - \eta_o)} = \text{constant, independent from } i.$$ (Eq. 3.68)

With $\eta_o = \dfrac{\Psi - \tau}{\Psi + 1}$ and $\tau = \left(\tau_{whole} \right)^{\frac{1}{n}}$ (Eq. 3.69)

the "total quality grade of the thermal ratio" has been found at the cascade with

$$\eta'_{cascade} = \left(\frac{\eta_o}{1 - \left(\tau_{whole} \right)^{-\frac{1}{n}} \cdot (1 - \eta_o)} \right)^n$$ (Eq. 3.70)

From this equation the following can be inferred:

- If the quality grade of the refrigerating capacity $\eta_o = 0$, so the saving also equals 0;
- if $\eta_o = 1$, so also $\eta' = 1$.

. if we yet assume that at a two-stage cascade a temperature ratio total of τ_{whole} is required, which cannot be realized with a single element (because of $\eta_0 = 0$), but for example with a binary product $\tau_{whole} = \tau_{whole}^{1/2} \cdot \tau_{whole}^{1/2}$ outlined with 2 single elements of which each realizes a temperature ratio $\tau_{whole}^{1/2}$ with a quality grade > 0, then the task is solved with this two-stage cascade.

As an example of the cascade effect for $n = 2$ it can be set $\tau = \Psi$ according to Equation 3.31 as an upper limit of the temperature ratio in case of a single thermocouple whereby according to Equation 3.46 the temperature ratio becomes at the single thermocouple $\eta_0 = 0$. With the temperature ratio $\tau_{whole}^{1/2} = \Psi^{1/2}$ at the single element it results according to Equation 3.46

$$\frac{\Psi - \left(\tau_{whole}\right)^{1/2}}{\Psi + 1} \qquad \text{(Eq. 3.71)}$$

a value > 0 because of $\Psi > 1$ (Equation 3.30) so that the two-stage cascade reaches generally a quality grade > 0. Through insertion of $\Psi = 1.1$ into the above example (see Tab. 3.1: Numerical Example) and $\tau_{whole} = \Psi^{1/2}$ for the single element it results now $\eta_0 = 0.024$ instead of $\eta_0 = 0$. Thus is obtained, through insertion of $n = 2$ with the value η_0 into Equation 3.70, the quality grade $\eta_{(n=2)} = 0.50$. This relation is described with the formula:

$$\eta'_{(n=2)} = \left(\frac{\eta_0}{1 - \Psi^{-1/2} \cdot \left(1 - \eta_0\right)}\right)^2 > 0 \qquad \text{(Eq. 3.72)}$$

Since for $n = 2$ the temperature ratio τ_{whole} has the threshold value Ψ (Equation 3.32), it has indeed achieved a qualitative improvement by the two-stage cascade design by reaching a positive heat pump effect with $\eta' > 0$ instead of $\eta' = 0$.

3.4.5 The Border Crossing Point $n \to \infty$

For a given temperature lift $\tau_{whole} = T_h/T_o$ of the entire cascade, the threshold value with η_0 must be determined according to Equation 3.71

$$\lim_{n \to \infty} \eta' = \lim_{n \to \infty} \left(\frac{\eta_0}{1 - \tau_{whole}^{-\frac{1}{n}} \cdot \left(1 - \eta_0\right)}\right)^n \qquad \text{(Eq. 3.73)}$$

The determination was based essentially on the elemental approximation

$$\left(\tau_{whole}\right)^{\frac{1}{n}} \approx 1 + \frac{1}{n} \cdot \ln \tau_{whole} \qquad \text{(Eq. 3.74)}$$

for large n, from which the further approximations result

$$\eta_0 \approx \frac{\Psi - 1 - \frac{1}{n} \cdot \ln \tau_{\text{whole}}}{\Psi + 1} \quad \text{and} \quad 1 - \eta_0 \approx \frac{2 + \frac{1}{n} \cdot \ln \tau_{\text{whole}}}{\Psi + 1} \qquad \text{(Eqs. 3.75, 3.76)}$$

Through insertion into Equation 3.73 and executing the border crossing point it comes to

$$\frac{\dot{Q}_o}{\dot{Q}_h} = \tau_{\text{whole}}^{-1} \cdot \eta'_{n=\infty} = \tau_{\text{whole}}^{-\frac{\Psi+1}{\Psi-1}} \qquad \text{(Eq. 3.77)}$$

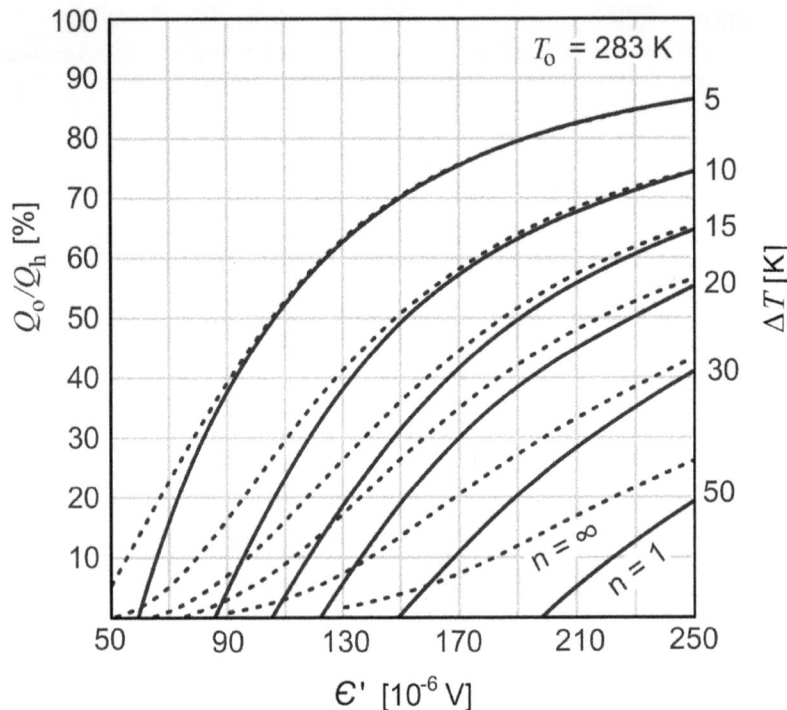

Fig. 3.5: Maximum saving on thermal heat

This saving is outlined in Figure 3.5 for T_o = 283 K (e. g. groundwater temperature) for a number of stages $n = 1$ with η' according to Equation 3.59 by the solid-line array of curves and for $n = \infty$ according to Equation 3.77 by the dotted array of curves. Therefore, for example, groundwater warming up to room temperature (i. e. by 20 K) with $\mathcal{E}'$ = 230 µV does not yet consume even half the amount of electric energy as with heating by Joule heat (the saving would be therefore larger than 50 %).

The effect for a large number of stages is shown by the asymptotic path of the curve for $n = \infty$ (dotted array of curves) so that also for infinitesimal thermoelectric powers still remains an effect > 0.

3.5   Summary of Results

In the introduction to Gehlhoff[22], which Altenkirch contributed, it says, 'Opposite to conventional cold production methods, it is enabled to produce cold without a working fluid by harnessing the Peltiér effect.' When the effective thermoelectric power in accordance with Equation 3.29 is defined

$$\mathcal{E}' = \frac{2\,\mathcal{E}}{\left(\rho_a^{1/2} + \rho_b^{1/2}\right)}$$

(Eq. 3.78)

where $\mathcal{E}$ is the actually measured thermoelectric power of the two components, and ρ_a, ρ_b the deviation of the components a, b from the Wiedemann-Franz-Lorenz Law (see Equation 3.25), so the maximal temperature decrease which is enabled by the electric current is given by $T - \theta_o$. Therein are

$$\frac{T_h}{\Theta_0} = \Psi = \frac{\Theta}{T_0} \quad \text{and} \quad \Psi = \sqrt{\left(1 + \frac{1}{4} \cdot \frac{\mathcal{E}'^2}{WFL}\right)} = \sqrt{1 + 10^7 \cdot \mathcal{E}'^2}$$

(Eqs. 3.79, 3.80)

where WFL is the Wiedemann-Franz-Lorenz constant ($WFL = 2.5 \cdot 10^{-8}$).

The efficiency Ψ is thus determined by the chosen couple of alloys and is a temperature-dependent material constant. The expense figure for the maximum of the refrigerating capacity—whose derivation has skipped here—is

$$\mathcal{E}^{-1}_{\max \dot{Q}_0} = \left(\frac{P}{\dot{Q}_0}\right)_{\max \dot{Q}_0} = 2 \cdot \tau_{\text{whole}} \cdot \frac{\Psi^2 - 1}{\Psi^2 - \tau_{\text{whole}}^2}$$

(Eq. 3.81)

The expense figure for the minimum expense of the wattage P for a desired refrigeration capacity $\dot{Q}_0$ with temperature T_0 is shown by

$$\mathcal{E}^{-1}_{\min P | \dot{Q}_0} = 2 \cdot \left(\tau_{\text{whole}} - 1\right) \cdot \frac{\Psi + 1}{\Psi - \tau_{\text{whole}}}$$

(Eq. 3.82)

If the thermocouples are designed in cascade such that the hot thermojunctions are cooled by the next row of cold thermojunctions, it is possible to reach any desired lowering of temperature and the smallest possible expense would be reached with an infinite number of stages

$$\mathcal{E}^{-1}_{\text{comb}}(\infty) = \left(\tau_{\text{whole}}\right)^{\frac{\Psi + 1}{\Psi - 1}} - 1$$

(Eqs. 3.83)

This expense of work is, however, extraordinarily great for large temperature differences—so it is, for example, according to Tab. 3.1: Numerical Example for the chosen thermocouple $\Psi = 1.1$, and thus the exponent in Equation 3.83 = 21, it means 50 K temperature increase starting from $T_0 = 283$ K up to $T_h = 333$ K $\Leftrightarrow T_h/T_0 = 1.176$ and $1.176 \cdot 21 = 29.47$ so that this convenient way of refrigeration only by means of electricity is out of question for large temperature differences.

For smaller temperature differences there were found sufficient good material combinations which led to practically usable refrigerating capacities and quality grades; over a long period, the material pair bismuth/antimony was considered as the combination with the greatest cooling effect. Improvements resulted by Gehlhoff's investigations[133]; recently, still more effective thermocouples have been found due to the availability of doped semiconducting materials.

3.6 Response and Present Developments

The response to Altenkirch's work was unsatisfactory. As Altenkirch reports in his autobiography[45], he received satisfaction many years hereafter when H. Lorenz expressed his admiration for his feat and confessed that he also had tackled the task without success. Also, E. Justi acknowledged Altenkirch's work in his treatise[74] as a "not yet surpassed theory"[73], [74].

In addition, in the recent relevant technical literature authors continued to invoke Altenkirch's theory as the first complete theory of the electrothermic refrigeration such as in L. v. Cube's edited handbook "Handbuch der Kältetechnik"[56] in the year 1997. The same applies to publications related with the following technical implementation: The junction of the positive and the negative leg of the cooling apparatus outlined in Figure 3.6 by a soldered-on copperplate means that there the thermoelectric powers $\mathcal{C}_a$ and $\mathcal{C}_b$ of both legs become effective against copper whereby the resulting thermoelectric power is given by $\mathcal{C} = \mathcal{C}_a + \mathcal{C}_b$ ($\mathcal{C}_a$: thermoelectric power of Bi_2Te_2 against Cu, $\mathcal{C}_b$: thermoelectric power of Cu against Bi, and $\mathcal{C}$: thermoelectric power of Bi_2Te_2 against Bi).

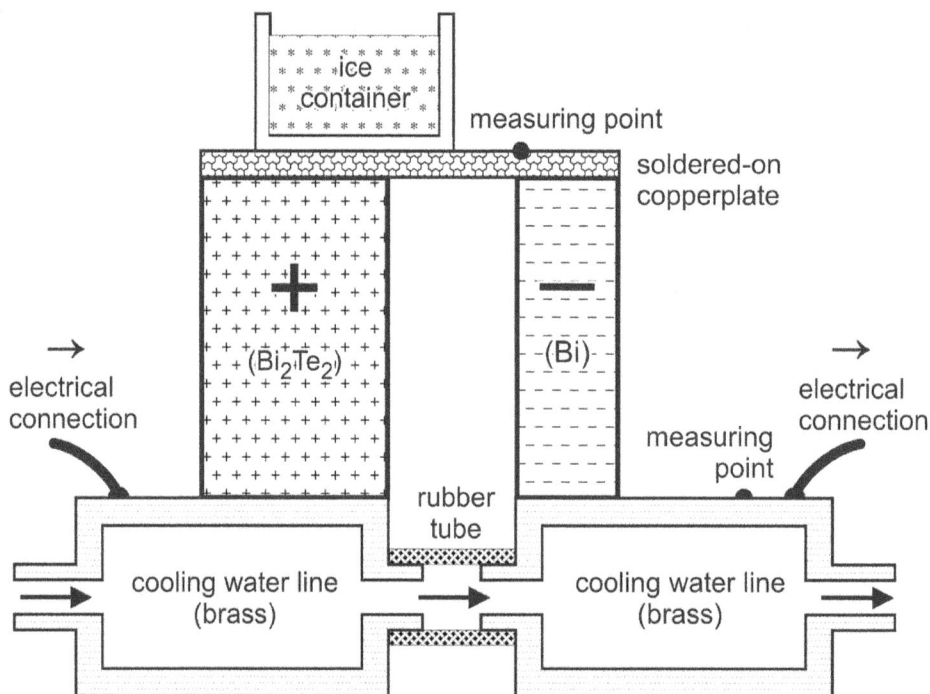

Fig. 3.6: Prototype of an electrothermic cooling apparatus, see Goldsmid, H. J. and R.W. Douglas (1954). Manufacturer: General Electric Co. Ltd., Wembley

That means the entire cooling effect is calculated as the sum of the signed Peltiér heat flows $\mathcal{C}_a T_o I$ and $\mathcal{C}_b T_o I$. The "summing up of the absolute values of the *cooling*" occurs in the copper plate due to the thermal conduction and is drawn from the ice canister. The mentioned thermojunctions on the cold side are on the same potential, but not those on the hot side.

Here, the thermoelectric powers of Bi_2Te_2 and Bi against brass come analogously into play. Due to the different potentials of the thermojunctions against the brass, an electrical insulated line segment (rubber tube) must be inserted into the connecting cooling water line. The summing up of the two Peltiér heat flows on the hot side is taken over by the cooling water.

3.6.1 Current Development

One example for a current development in the field of the thermoelectric production of energy, which was treated by Altenkirch as well (see Section 11), has been provided by the automobile manufacturer BMW[24], who intended to install so-called 'thermoelectric generators' in cars powered by the temperature difference between waste gas and external air in order to provide relief to the electrical car system—a (somewhat "broad-brush") quality criterion for the assessment of the suitability of thermoelectric materials is there stated

$$ZT = S^2 \cdot \sigma \cdot \frac{T}{\kappa}$$

(Eq. 3.84)

with: $S = \mathcal{C}$ (thermoelectric power)
σ = electric conductivity
κ = heat conductance
T = absolute temperature

Compared with the square of Equation 3.33 one will notice rather an oversimplification.

Effects at a Two-Stage Cascade

The thermojunctions of the cascade outlined in Figure 3.7 have upward and downward contact on level 3. The first copper plate in position S1 on level 3 at the right end of the cascade acts as an electric current supply (positive pole). The upper stage of the cascade with the expense figure ζ_1 has longer legs; therefore the electric current is lower because of the higher electric resistance as in the lower cascade where the legs are shorter.

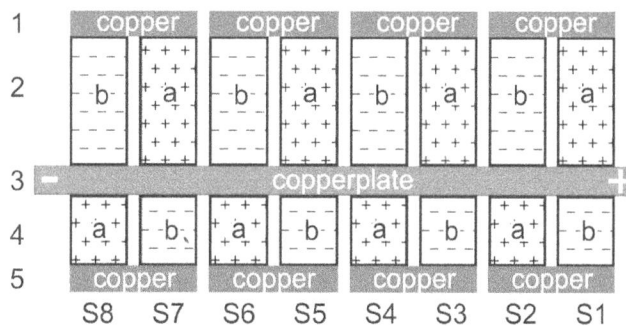

Fig. 3.7: Two-stage cascade in a planar arrangement according to Joffe[71]

The cold $Q_{o\ stage\ 1}$ originates at the upper thermojunctions on level 1 due to the passage of the electric current from a to b. The upper thermojunctions on level 3

---

[24] VDI-Nachrichten, Düsseldorf, October 30, 2009, wop[173]

connect into the direction of the electric current at first copper with material a and afterward material b with copper. We have therefore the contact sequence[25]:

$$Cu \| a, b \| Cu \text{ or } b \| Cu \; Cu \| a = b \| a$$

Thus, the heat originates here at the hot thermojunction of the upper stage, but in two portions at $Cu \| a$ and $b \| Cu$ which yield only summed up the amount of heat $Q_{\text{stage }1} = (1+\zeta_1) \cdot Q_{\text{o stage }1}$.

The lower thermojunctions on level 3 first connect into the direction of the electric current the copper with material b and then the material a with copper. We have therefore the contact sequence total: $Cu \| b, a \| Cu$ or $a \| Cu \; Cu \| b = a \| b$.

Thus, the cold originates here at the cold thermojunction on the lower stage, but again in two portions on $a \| Cu$ and $Cu \| b$, which again yield only summed up the amount of cold $Q_{\text{o stage }2}$ that has to compensate the amount of cold $Q_{\text{stage }1}$:

$$Q_{\text{o stage }2} = (1+\zeta_1) \cdot Q_{\text{o stage }1} \qquad \text{(Eq. 3.85)}$$

In the same ratio $(1+\zeta_1)$ the electric current on stage 2 is larger than on stage 1. Also in this ratio the leg length on stage 2 is shorter than on stage 1. The cooling load to be transferred to the water as cooling medium on level 5 then is

$$Q_{\text{stage }2} = (1+\zeta_1) \cdot (1+\zeta_2) \cdot Q_{\text{o stage }1} \qquad \text{(Eq. 3.86)}$$

The heat transfer of the mentioned parts of the absolute values of heat in horizontal direction, which also can have opposite signs, is also enabled between the columns of the legs 2 and 3, 4 and 5 as well as between 6 and 7 via the copper plate on level 3 in longitudinal direction; likewise from there in vertical direction upward and downward. At the margins it is however only enabled with appliances closed in a ring! In one example is achieved a by 27.5 % larger COP of cooling for the combination of a p-type Bi_2Te_3-Sb_2Te_3 alloy with a n-type Bi_2Te_3-Sb_2Se_3 alloy compared to a single element with 0.315 at $T = 273 + 40$ K and $T_o = 273 - 10$ K. Thus a maximum temperature lift is already achieved by 74 K. With the combination of extremely thin semiconducting layers of bismuth tellurides (Bi_2Te_3) and antimony tellurides (Sb_2Te_3), the team around Venkatasubramanian (et al.) (see Nature 11/2001),

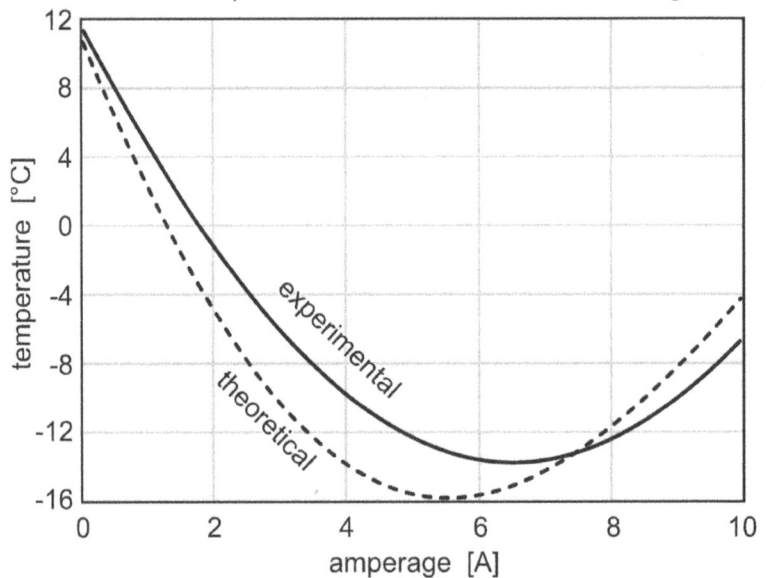

Fig. 3.8: Experimental versus theoretical temperature profile at the cold thermojunction

Research Triangle Institute, North Carolina (U.S.A.) achieved a significant increase of the

---

[25] "$\|$" symbolizes the electric contact, that means, by thermojunction [solder junction]; "Cu" denotes the electric connection by a copper conductor

effective thermoelectric power (see Tab. 3.1: Numerical Example). With more than two stages, horizontal layers must be inserted which are electrically insulating but well-conducting the heat. Joffe developed a well-qualified substance for this purpose, a blend made of silicone varnish and metal powder (see below).

Three-stage Cascade

Figure 3.9 shows according to a patent by Hans Voigt[119] a three-stage cascade in which the heat transfer has accomplished by well-conducting metallic joints made of copper which are simultaneously integral parts of the thermojunctions and are joints between principal and auxiliary legs of different thermoelectric power. The occurring Peltiér effects cover the desired thermal coupling of the stages whereby the thermal balance is achieved through adapting the electric currents and the cable cross-sections (the width of the legs in the figure) so that

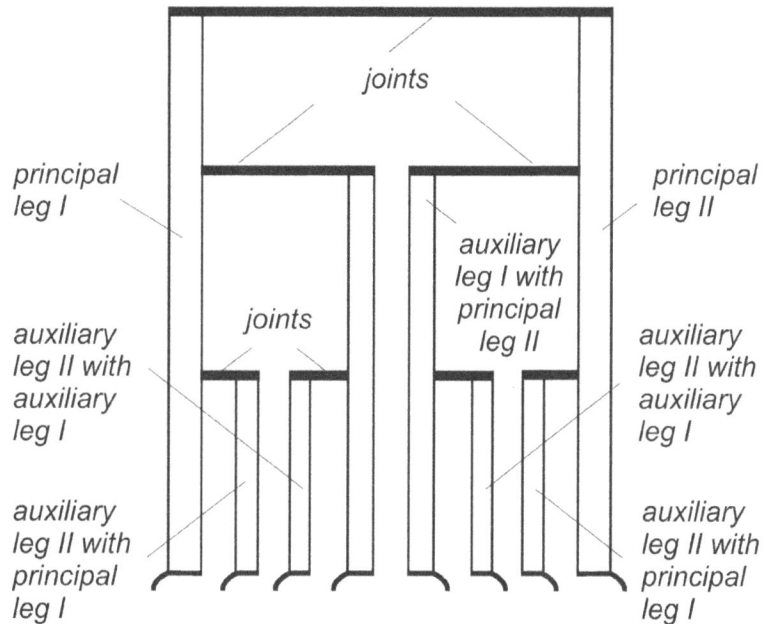

principal leg I

auxiliary leg II with auxiliary leg I

auxiliary leg II with principal leg I

joints

joints

auxiliary leg I with principal leg II

principal leg II

auxiliary leg II with auxiliary leg I

auxiliary leg II with principal leg I

Fig. 3.9: Three-stage cascade acc. to H. Voigt [167]

the same time electrical insulating but thermally conductive layers are not needed—the buildup of a three-stage cascade according to Joffe[71] shows Figure 3.10.

3.6.2 Some Specific Practical Problems

· **The contact resistances at the thermojunctions**— improvement with interposition of a sheet made of copper that can simultaneously serve as a heat transfer area plus application of solder alloys (51 % indium and 49 % tin) with a low melting point at 117 °C (see the just mentioned paper by Joffe).

· **The decrease of the thermal resistances at the electrically insulating intermediate layers** by heat transfer layers between successive stages of one cascade—according to Joffe improvement with silicone varnish with 6 % aluminum powder content; nowadays there are thermal conductance pastes on offer from different companies.

· In "off" position the whole aggregate has unfortunately the effect being a thermal bridge between "cold" and "hot."

Fig. 3.10: Three-stage cooling aggregate acc. to Joffe, A. F. et al. (1956)

4 Reversible Heating

4.1 Reversible augmentation of heat by "heat transformation" (as of 1913)

Altenkirch acquired the fundamental theory for this purpose by studying Gustav Anton Zeuner's[168] handbook on technical thermodynamics and especially the technical journal "Zeitschrift für die gesamte Kälteindustrie," edited by Hans Lorenz[170].

Fig. 4.1: Geh. Rat [privy councilor] Prof. Dr. HANS LORENZ, Danzig (today Gdańsk, Poland), Chairman of the German Refrigeration Association.

Fig. 4.2: Geh. Bergrat [privy councilor of the mining industry] Gustav ANTON ZEUNER (1828-1907): Founder of the scientific school "Technische Thermodynamik" (TU Dresden, custodies. Photograph: TU Dresden, AvMZ (Archiv))

Altenkirch described in his autobiography an early recognition of a scientific interest in the words[45], 'Studying the second law of thermodynamics had opened my eyes to the huge practical implications reversible heat generation must acquire if economically viable ways of implementing it could be pursued. This heat transformation also opened up new methods to improve the production of work from heat of high temperature or with utilization of low outdoor temperatures.'[26]

After Altenkirch had to abandon the connections developed with the thermocouple due to too low an order of magnitude of the effective thermoelectric powers, he

---

[26] Under reversible heating, cooling, reversible heat exchange, and mass exchange Altenkirch understood exchange processes of which driving forces, that means for instance temperature differences or partial pressure differences could be diminished due to appropriate measurements (see Figure 5.4, and Figure 5.5 (as a counter example), solution rerouteing in the absorber/desorber, Section 5.6) as well as the avoidance of typical irreversible processes, for instance with the heating (heating with the Peltiér effect or due to a heat pump instead of electrical resistance heating), the heating due to heat transfer from heating media, the mass transfer in counter current to thermic or hygroscopic layered solid exchangers (regenerators), and much more.
For the assessment of the losses of the thermic insulated sub-systems the entropy balance of the in- and outgoing currents are used.

searched for alternate methods to develop an economical exploitation of reversible heat generation. Thereby, he also took into account an energy-generating sub-system (as in the case of the thermocouple the generator for the electric energy). He was especially interested in *simple* methods. His understanding of "simplicity" above all meant the lack of mechanically moving parts. With this "simplicity" a long operational life span and a low maintenance should be guarantied, and the building materials should be inexpensive.

4.1.1 Reversible Heating due to Power-Heat Coupling [Cogeneration] with Piston and Lifting Elements

As a first solution came into question—even though working with mechanically moving parts in the shape of piston and lifting elements—the compression refrigeration machine as a heat pump in connection with a steam power machine (as an energy-generating working machine).

Consequently, as early as 1912 he pursued the path of power-heat coupling [cogeneration] for industrial facilities and municipal heating systems which became common technological knowledge in the recent past:

To this end, he published the papers:

- "Umkehrbare Heizung"[4]. [Reversible Heating]
- "Die Erhöhung der Wirtschaftlichkeit von Heizungsanlagen"[7]. [Increase of the Efficiency of Heating Facilities]
- "Reversible Wärmeerzeugung"[12]. [Reversible Generation of Heat]

Based on Sir William Thomson's (Lord Kelvin, 1824–1907) coupling of power-heat generation and cold production [cogeneration][114] for space heating suggested in 1852, as is outlined in Figure 4.3, Altenkirch proposed an essential improvement of this connection:

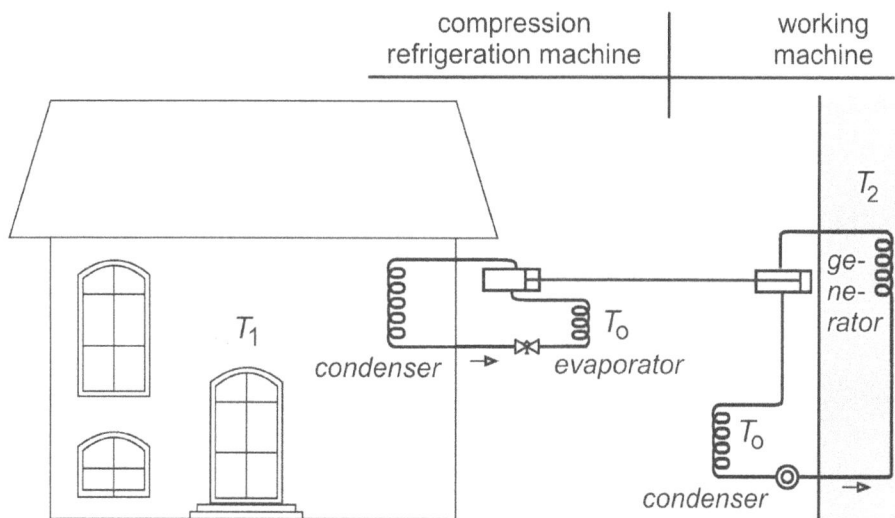

Fig. 4.3: Power-heat coupling [cogeneration] acc. to Thomson

Whereas with Thomson's connection the waste heat of the working machine was discharged to the cooling medium (ground water/cooling water) at temperature

T_o, Altenkirch achieved a higher economical feasibility by supplying, in addition, the liquefaction heat of the refrigeration machine together with the waste heat of the working machine (which was released in its condenser) immediately to the heating at temperature T_1 (see Figure 4.4). In the original connection, compared with this, (a smaller) partial amount of utilized thermal heat increased by the lower liquefaction temperature at T_o first resulted from the amount of work of the working machine via the detour over the heat pump.

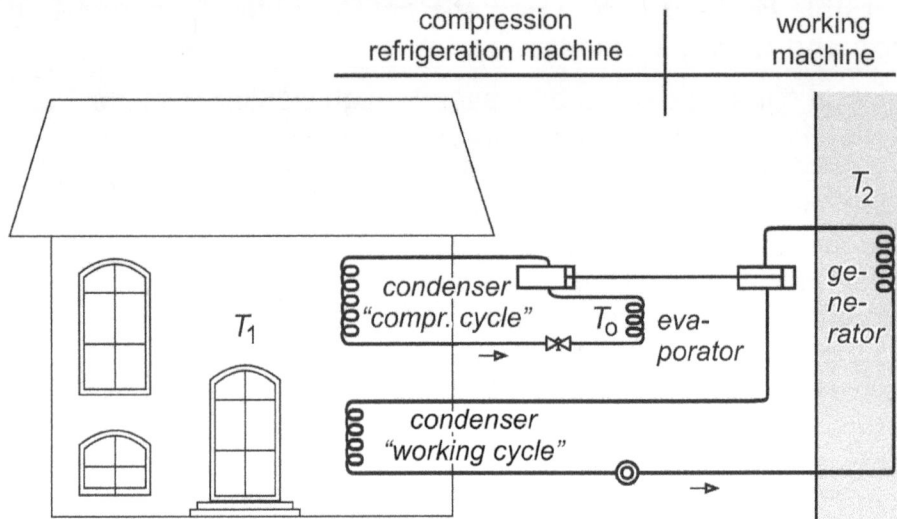

Fig. 4.4: Power/heat/cold coupling [trigeneration] according to Altenkirch

Thus, with Altenkirch's connection outlined in Figure 4.4 both the liquefaction heat of the working machine and the liquefaction heat of the compression refrigeration machine were used for the space heating at temperature T_1.

Thereby, the refrigeration machine/heat pump system took an absolute value of heat from the ground water or waste water in its evaporator at temperature T_o, which, increased by the supplied work in the compressor of the refrigeration machine, was released as thermal heat in the condenser. The compression work was generated from the working machine and transmitted via rods to the heat pump whereby the work evolved, as is known, due to the temperature gradient $T_2 - T_1$ between the thermal heat at T_2 in the generator of the working machine, and the released heat, used as spatial thermal heat, in the condenser of the working machine at T_1.

Altenkirch: [The connection in Figure 4.3] 'would have been less successful since in both connections the working machine and the refrigeration machine were working with unavoidable losses': The improvement of the connection according to Figure 4.4 was evident because even the waste heat of the working machine increased with the losses of the working machine[160], and the loss of transmission of mechanical work from the working machine to the heat pump was the smaller, the smaller the transmitted absolute value of work was itself.

Lord Kelvin's proposal was far ahead of its time but had already predecessors even though the full insight into the range of consequences at that state of thermodynamics had to be missed. We are talking about the special field of devices for distillation or concentration of liquids in vacuum.

Already in the year 1834 the principle of heat pump[96] was published. The heat supply to the liquid for the vapor expulsion is largely taken over by the condensation heat of the compressed vapor itself which is here caused by a steam jet.

The proportions of the here discussed connections are again illustrated in the T-S diagram in Figure 4.5: Herein the (nearly square) surface between the line segment from S_1 to S_2 on the abscissa and from T_0 to T_1 on the ordinate is to be considered: Again is evident that with Thomson's connection a greater absolute value of work has to be generated from the steam engine first by the hatched area after it causes an enlargement of the pumped-up heat via the rods by the heat pump while the waste heat (surface between the ordinate T_0 and the line segment from S_1 to S_2 on the abscissa) flows unused to the cooling medium at T_0.

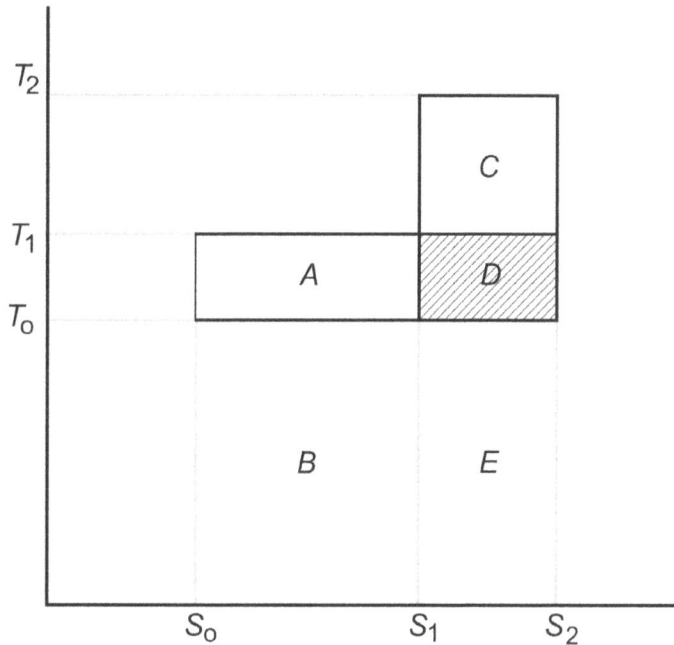

Fig. 4.5: T-S diagram of the connections according to Thomson and Altenkirch

According to Altenkirch the waste heat of the heat engine (surface between the ordinate T_1 and the line segment S_1, S_2) is directly supplied (at T_1) to the heating so that the heat pump only has to deliver the corresponding absolute value of heat between the ordinate T_1 and the line segment S_0, S_1.

Summary

<u>Again summarized as to the steam engine:</u>

- With Thomson's connection the whole transmitted work of the steam engine to the heat pump is constituted by the sum of the surfaces C and D as well as the waste heat at T_0 by the surface E.
- In contrast, according to Altenkirch the transmitted work of the steam engine is only equivalent to the surface C and the share of the transmitted heat at T_1 is constituted as utilizable waste heat by the sum of the surfaces D and E.

<u>Again summarized as to the heat pump:</u>

- With Thomson's connection the heat taken from the heat reservoir at T_0 is constituted by the sum of the surfaces B and E and the supplied work by the sum of the surfaces A and D.
- With Altenkirch's connection the heat taken from the heat reservoir at T_0 is only constituted by the surface B and the supplied work by the surface A.

Again as to the *coupling* of steam engine and heat pump:
- With Thomson's connection C and D is transmitted to A and E without a remainder, that means: $C + D = A + E$ is valid.
- With Altenkirch's connection only D is transmitted to A; compared to Thomson's connection, the losses of transmission due to this measure with the same summarized heat output ($= A + D$) are smaller since less work has to be transmitted. The other part of the utilizable thermal heat provides the "waste heat" of the steam engine.

Together with his former private student, Tenckhoff, Altenkirch developed the knowledge of the technical principles and wrote them down in a Patent[160].

4.2   Cascade Design

To this end, Altenkirch commented, 'In many cases, where otherwise a reversible heat generation is desired, a large temperature difference cannot be avoided when, for instance, water vapor of high tension shall be generated whose waste heat is not utilizable for whatever reason. It is obvious that for such facilities it is always more difficult to achieve an economical feasibility the higher the temperature difference is. For, on one hand, the expense of machinery is getting higher; and due to it the expense of work rises so that the saving is diminished compared to the irreversible heating.

One could have thought that with large temperature differences the losses are even so high that the savings are almost disappearing. Several refrigeration machines have to be connected in cascade, each has its special quality grade and if, these quality grades, say, have to be multiplied with one another in order to obtain the quality grade total, like it had been done with the series connection of mechanical engines, so the power demand would be so high to overcome temperature differences by, for example, 150 °C so that the saving would be only slightly.

This is, however, not the case, for two reasons, namely: firstly the quality grade for the heat generation is more favorable than for the refrigeration. Less for the fact that irradiations did not appear as losses anymore (emission losses of the condenser occurred instead)—but in simple consequence of the First Law since the wattage P was added up to the refrigerating capacity $\dot{Q}_o$ and it resulted the heat wattage $\dot{Q}_h = P + \dot{Q}_o$':

Starting from the definitions (Equations 4.1, 4.2) of the Coefficients of Performance of refrigeration COP_0 and (Equation 4.2) heat generation COP_h

$$COP_o = \frac{\dot{Q}_o}{P} \text{ and } COP_h = \frac{\dot{Q}_h}{P} \qquad \text{(Eqs. 4.1, 4.2)}$$

that means, from the yield of cooling or heating, respectively, of an expense of work (wattage) P as well as from the definitions of the quality grades (see Equation 3.46) become

$$\eta_o = \frac{\dot{Q}_o}{P} \cdot (\tau - 1), \tau^{-1} = \frac{T_o}{T_h} \text{ as well as } \frac{\dot{Q}_h}{P} \cdot (1 - \tau^{-1}) \qquad \text{(Eq. 4.3)}$$

Herewith, Altenkirch generates, like with the thermocouple in Section 3.4, a "quality grade η' of the heat ratio" $\dot{Q}_o/\dot{Q}_h$ [27] on a heat pump by comparing it with the theoretical value τ^{-1} of the heat ratio. The result (see Equation 3.57) is shown as

$$\frac{\dot{Q}_o}{\dot{Q}_h}=\eta'\cdot\frac{\dot{Q}_o}{T_h}=\frac{\eta_o}{\eta_h}\cdot\tau^{-1}=\frac{\eta_o}{1-\tau^{-1}\cdot(1-\eta_o)}\,\tau^{-1} \qquad \text{(Eq. 4.4)}$$

The Equation $\dot{Q}_o/\dot{Q}_h=\eta'\cdot\tau^{-1}$ describes the saving on needed work opposite to irreversible heating with JOULE heat. Through evaluation of Equation 4.4 results eventually the formula for the quality grade η' of the heat pump

$$\eta'=\frac{\eta_o}{1-(\tau)^{-\frac{1}{n}}\cdot(1-\eta_o)} \qquad \text{(Eq. 4.5)}$$

The definition of η' is particularly appropriate for the treatment of a cascade out of single heat pumps $(Q_j;...)$ from $j=0$ to k heat pumps whereby the heat flow Q_{j+1} inside the cascade is produced by the transfer of the heat flow Q_j from one stage j to the following.

The quality grade of one cascade with k stages follows from Equation 3.70 —as the product of the quality grades of the heat ratio of the several stages, likewise, of course, the temperature ratio adjacent to the cascade

$$\tau_{whole}(k)=\Pi_{i=1...k}\,\tau_i$$

is the product of the several temperature ratios adjacent to the several stages. When all quality grades and temperature ratios are equal, it results for k stages

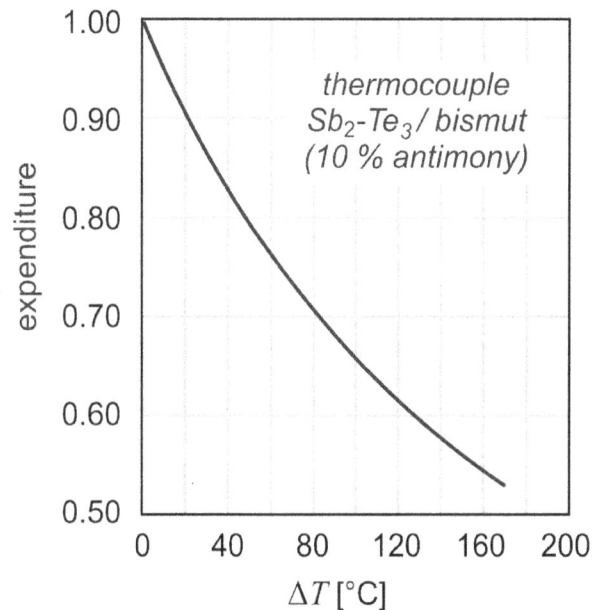

Fig. 4.6: Saving on reversible compared to irreversible heating in thermoelectric refrigeration

$$\eta'(k)=\left(\frac{\eta_o}{1-\tau_{whole}^{-\frac{k}{k}}\cdot(1-\eta_o)}\right)^k \qquad \text{(Eq. 4.6)}$$

From the heat ratio the expense of wattage P can also be *determined*. With (Equation 4.4) becomes

---

[27] Also see Equation 3.65 in Section 3 "Thermocouple for Reversible Heating" at: "Cascade Design."

$$\frac{\dot{Q}_h^{(k)}}{\dot{Q}_o^{(o)}} = \frac{\tau_G(k)}{\eta'(k)} \rightarrow \frac{P(k) = (\dot{Q}_h^{(k)} - \dot{Q}_o^{(o)})}{\dot{Q}_o^{(o)}} = \frac{\tau_G(k)}{\eta'(k)} - 1 \qquad \text{(Eq. 4.7)}$$

For a temperature ratio with $k = 3$ stages of $\tau_G(3) = 1.37$ which is equivalent to a temperature difference of ca. 100 K, and $\eta_o = 0.7$ according to Equation 4.6 becomes $\eta'(k) = 0.882$, and thus the expense of wattage $P(3)/\dot{Q}_{oj}^{(1)} = 0.553$. In the theoretically best case is $P(3)/\dot{Q}_{oj}^{(1)} = \theta - 1 = 0.37$.

'The ratio of the favorable power consumption to the real power consumption is therefore $0.37/0.55 = 0.67$, thus only a little worse than with a single refrigeration machine which is working between small temperature differences. This result was of general interest. With this for the cascade design favorable result successes can be expected for reversible heat generation with larger temperature differences.'

It can be concluded that at least with one kilowatt hour with reversible heat generation with 3 stages $1/\zeta_{id} = 1/(\theta - 1) = 1/0.37 = 2.71$ almost three times more heat can be performed than with irreversible heat generation. The reversible water vapor production of electric energy from the water powers of Switzerland and Scandinavia is therefore only a question of economical feasibility.

The expected savings due to reversible heating according to Equation 4.4 (one-staged with a quality grade of $\eta = 0.7$) for distinct temperature differences compared to irreversible heating is outlined graphically in Figure 4.7; in Figure 4.6 an achieved saving at least with one thermocouple (applicable to distillation processes with little temperature differences) is shown.

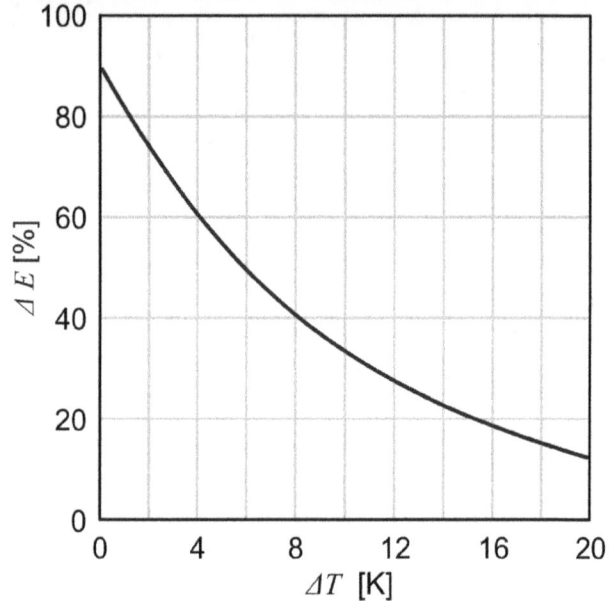

Fig 4.7: Savings due to reversible heating (one-stage, with $\eta = 0.7$)

4.3 Response to the Work Dealing With the Connections According to 4.1

Also, these pioneering pieces of work Altenkirch carried out with his former student, Bernhardt Tenckhoff, in the year 1912/13, when, as mentioned earlier, his salary for private lessons served for financing his university studies. The results of this research he presented in a lecture at the International Congress of Refrigeration in Chicago[3] in 1913.

These solutions attracted internationally a great deal of attention. So the renowned Russian refrigeration engineer Ryazantsev was motivated by Altenkirch's lecture in Chicago to pursue further calculations—but their solutions lay still too close to the thresholds of economical feasibility.

After World War II, the practical implementation of Altenkirch's work led to a contract with the company Borsig, on one hand, and the companies Zölly and Escher-Wyss[28], on the other hand; and later with the company Siemens-Schuckert-Werke. He also worked on projects, but the hyper-inflationary period of the Weimar Republic was inappropriate for farsighted investments.

Altenkirch therefore searched for devices with which he could carry out the desired heat transformation by simpler means, without visible mechanical motion and with a minimum of moving parts. Thus he also considered the steam jet ejector chiller:

4.3.1 Reversible Heat Generation with the Steam Jet Ejector Chiller

During the First World War Edmund Altenkirch had to teach extended lessons in mathematics and physics at the high school (Realgymnasium[29]) in Berlin-Karlshorst[30] as a Hilfsdienstpflichtiger[31]. Besides his school teaching, he made measurements on the steam jet ejector chiller at the Technische Hochschule[32] in Berlin-Charlottenburg[30], a device invented by Josse and Gensecke. The instance was the test for the possible use of the steam jet ejector chiller for reversible heating related with his Patents[132], [160].

Therefore, Altenkirch made the quest for a heat transformer that was capable of fulfilling these qualifying criteria, that is, work with a minimum of moving parts (see the following Section).

---

[28] Swiss manufacture, inter alia for Hot and Cold Air Machines

[29] German academic high school with scientific subjects and modern languages (until 1937) [translator's note].

[30] A district of Berlin, capital of Germany [translator's note].

[31] In the First World War every male person who was not able to serve in the German army (Altenkirch had lost his left arm in a tragic accident—see Section 2, Paragraph 1) was deployed by law for charitable work. [translator's note].

[32] In German, this term is close to "technical university" [translator's note].

5 Absorption Machine as a Heat Transformer

5.1 Process Inside the Absorption Machine

In a binary mixture (like ammonia and water) we understand the term "absorption" as the uptake of a gaseous phase (refrigerant, e. g. ammonia gas) by a liquid mixture of a refrigerant and the solvent (water).

The heat to be emitted by this process is the difference between the enthalpy of the system (solution and gaseous phase) at the terminal and initial stage of the process, assuming it is isobaric. These stages describe spatial and temporal states of the process. The enthalpy is an extensive property which therefore depends only on the mass fractions of the substances, the content of the solvent and working medium, and the pressure and temperatures of the respective thermodynamic states.

The several sorption heat quantities are transferred from/to external media according to the mode of application of the absorption machine. This transfer occurs via separating walls as a heat transfer driven by the temperature gradient. For the *intensity* of both the heat transfer and the absorption (resorption) of the gaseous phase from the fluid solution the surface development and the average thickness of the sorptive liquid film is dependent—besides the temperature and concentration gradient—on the latter. When the gaseous refrigerant is situated as a component in a carrier gas, so, in addition, the mutual diffusion speeds of the gases into each other are decisive.

This description initially refers to the absorber (resorber of the resorption machine, see Figure 5.8). In the same way, only vice versa, the process takes place in the generator (degasser of the resorption machine): here it comprises the desorption or expulsion heat which is supplied to the liquid solution to release the vapor or gas. Due to the surface tension of the liquid, the processes are however not exactly mirror-inverted in detail since for example the development of gas bubbles in the liquid has no equivalent in the case of absorption.

Later these processes will be considered here idealized especially as parallel and linear proceeding changes of temperature and concentration in liquid and gas along the horizontal dimensions visible in the Figures 5.1 and 5.2 (and vertical dimensions in Section 7, respectively) of the sorption apparatuses.

5.2 Altenkirch's Representation of Carré's Absorption Machine

In 1914 Altenkirch became aware of Carré's absorption refrigeration machine and thus reached the refrigeration technology in the narrow sense although from a more general perspective[33]. In this field, he achieved his most important technical feats. The starting point for his consideration was Carré's absorption refrigeration machine as it is outlined in Figure 5.1 as a schematic.

---

[33] To this end, see his papers: "Reversible Absorptionsmaschinen" [Reversible Absorption Machines] 1913 and 1914 [5]; "Die Erhöhung der Wirtschaftlichkeit von Heizungsanlagen durch den Einbau von Kältemaschinen" [The Increase of the Economic Efficiency of Heating Facilities due to the Installation of Refrigeration Machines][7]; "Absorptionskältemaschinen" [Absorption Refrigeration Machines], Publisher: VEB Verlag Technik, Berlin in 1954, unfinished[44]; "Absorptionskältemaschine zur kontinuierlichen Erzeugung von Kälte und Wärme oder auch von Arbeit" [Absorption Refrigeration Machine for Continuous Generation of Cooling and Heating or Work][160]

According to Altenkirch's documentation, both the vessels and the connecting lines are arranged almost in a two-dimensional coordinate system so that the image ordinate corresponds to the logarithm of the prevailing pressure and the abscissa to the negative of the reciprocal absolute temperature.

In a log p-1/T diagram the curves of the constant working medium concentration according to the equation of Clausius and Clapeyron[34], essentially based on the only slightly temperature-dependent evaporation heat of the working medium from the solution, are approximately straight lines (see for example the solution plot in Figure 5.2). It deals with the following sub-processes:

· In the generator of Fig. 5.1 the working medium is released from the rich working medium solution which enters in an upward direction, and is conducted as vapor into the condenser whereby the thermal heat $\dot{Q}_h$ is transferred by a stylized coil in reverse U-shape.

· The vaporized working medium travels through the joining horizontal line segment (displayed in Figure 5.1 without shading) into the condenser; the fluid working medium is conveyed farther to the regulating valve V_1 in which it is expanded from the high pressure p of the condenser/expeller[35] to the low pressure p_0 of the evaporator/absorber and then again is converted to the vaporized state in the evaporator with emitting the cooling $\dot{Q}_0$ which is caused by aspiration of the vapor from the cold and poor solution in the absorber.

· Condenser and evaporator are stylized without liquid compartments.

· The (so-called rich) solution enriched with working medium in the absorber is again re-compressed in the solution pump (located bottom left next to the absorber) up to the high pressure p and is then fed through the outgoing line from the pump into the cold side (in the picture on the left hand side) of the generator.

· After degasification of the working medium in the generator, the solution which is now *poor* in working medium leaves the generator (in the picture on the right hand side) and is farther precooled inside the inner tube of the heat exchanger (stylized as a double pipe) with the rich solution (as previously mentioned) from the absorber.

· In the regulating valve V_2 it is expanded to the pressure p_0 of the absorber and fed into its warm side (outlined in the picture on the right hand side).

---

[34] Named for its originators: Rudolf Clausius (1822-1888) and Emile Clapeyron (1799-1864). The equation is derived (with S'', S': molar entropy, V'''; V': molar vapor and liquid volume) from

$$\frac{dp}{dT} = \frac{S''-S'}{V''-V'} \qquad \text{(Eq. 5.1)}$$

By introducing the molar heat of the evaporation q and the universal gas constant $R = 8.31$ J K^{-1} mol^{-1} becomes

$$\frac{d \ln p}{d \, 1/T} = \frac{-q}{R} \qquad \text{(Eq. 5.2)}$$

With the molecular weight of the substance the heat of evaporation r is written

$$r = -\frac{19.13}{\mu} \cdot \frac{d \log p}{d \frac{1}{T}} = -\frac{19.13}{\mu} \cdot \frac{\Delta \log p}{\Delta \frac{1}{T}} \qquad \text{(Eq. 5.3)}$$

---

[35] In the following text the term "generator" stands for "expeller" (see Nomenclature)

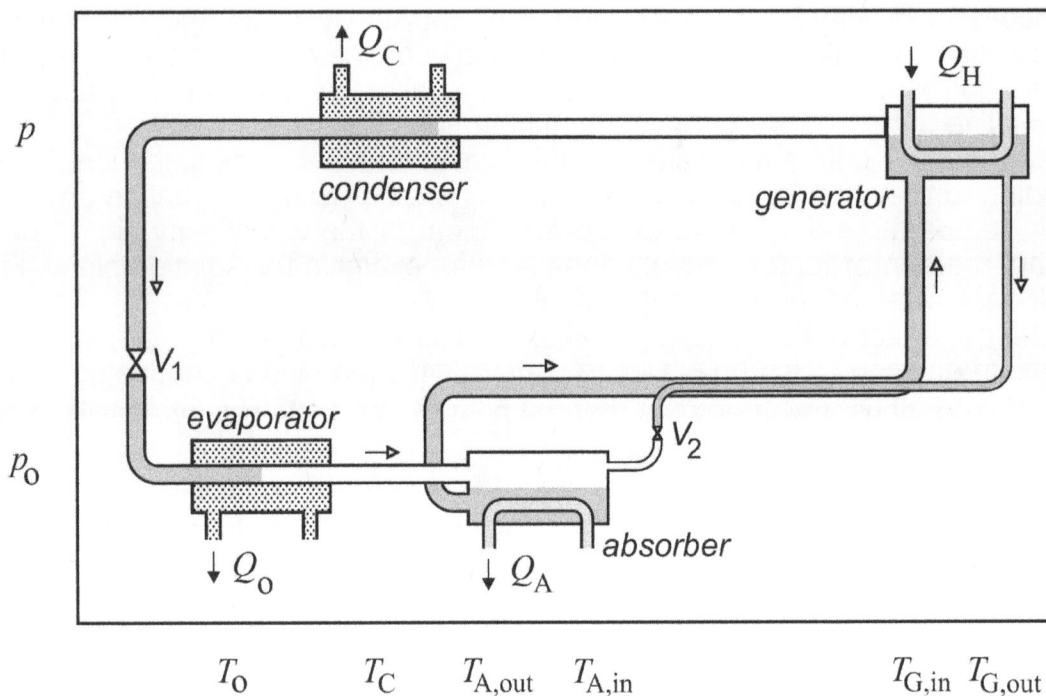

Fig. 5.1: Schematic of the absorption machine according to Carré

. As mentioned in the Introduction, he arrived at his innovation on this type of machine, as outlined in the following sections (from Section 5.3 onwards), through his conviction—gained by studying the second law of thermodynamics—that the

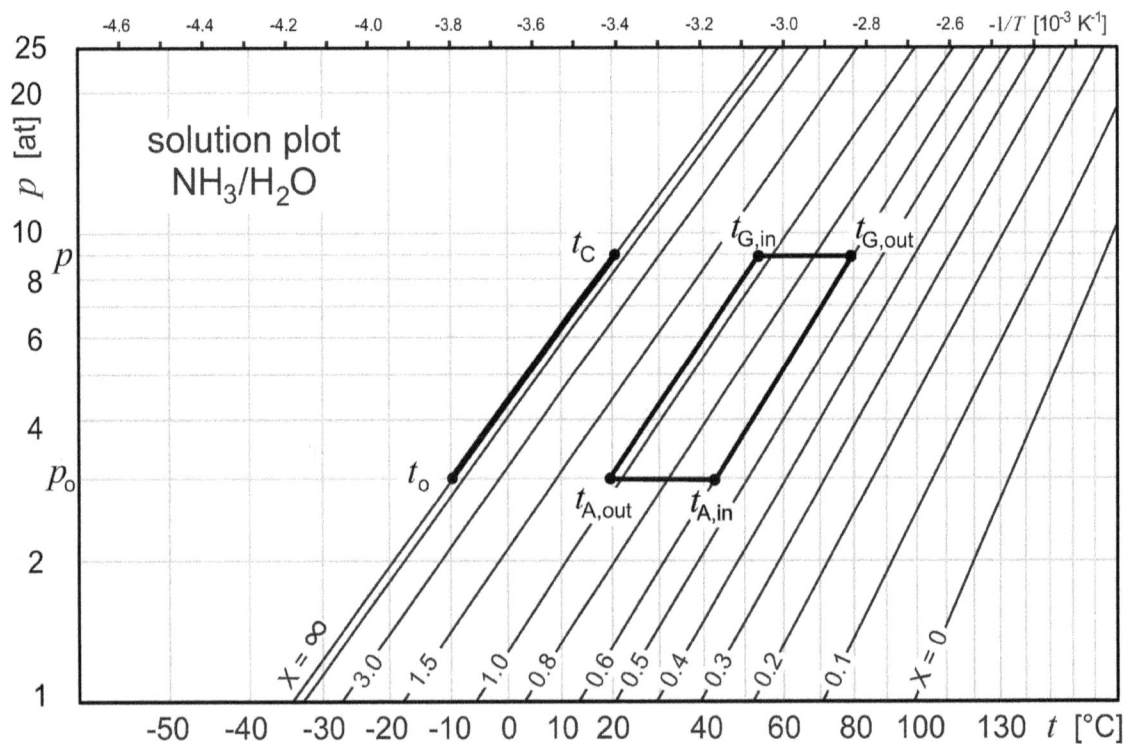

Fig 5.2: Working diagram of the absorption machine for the process acc. to Carré

theoretically possible effects in accordance with Equation 5.9 should also be obtainable in practice.

In Figure 5.2 the sorption processes run along the changes of state stylized with horizontal line segments, quasi as expulsion from $T_{G,inlet}$ to $T_{G,outlet}$ and absorption from $T_{A,inlet}$ to $T_{A,outlet}$ The result can be inferred from the concentration parameters of the curves.

The change of state of the working medium vapor produced in the generator runs on the isobaric line segment from $T_{G,outlet}$ to T_C where it is liquefied. The change of state of the condensate follows the saturation line of pure ammonia, it is expanded in the regulating valve V_1 (see Figure 5.1), and runs on the horizontal ("isobaric") line segment from T_o to $T_{A,inlet}$ to the absorber.

5.3 Absorption Machine and Power-Heat Coupling [Cogeneration] —Analogies

One subunit, consisting of the absorber and generator of the absorption machine plus the connection lines and the temperature exchanger of the solution cycle, is occasionally called a "thermal compressor."

Indeed, this subsystem of the absorption machine aspirates the working medium vapor from the evaporator and compresses it to the condensation pressure similar to the condenser of a compression refrigeration machine.

Compared to the lower temperature level T_A of the absorber determined by the thermodynamic mean temperature T_{Am}, the compression work for this process needed is called the *exergy* of the thermal heat emitted to the generator with the thermodynamic mean temperature T_{Gm}. This exergy (internal work) is constituted by the Carnot factor η_{Carnot}, formed by the thermodynamic mean temperatures (cf. Equation 5.9):

$$\eta_{Carnot} = \frac{T_{Gm} - T_{Am}}{T_{Gm}}$$ (Eq. 5.4)

Due to the running process in the absorption machine the heat of the lower temperature absorbed in the evaporator is raised from a lower temperature level to a higher one. Depending on the set task, the intended goal can be either the removal of heat at low temperature (for refrigeration) or the production of heat in the condenser C (see Figure 5.2) at higher temperature to generate thermal heat (heat pump). Therefore, in the following, it is generally possible to describe Altenkirch's innovations which are developed under the term heat transformation by envisaging the example of refrigeration as one field of application[36].

Figure 5.2 shows the NH_3/H_2O solution plot. On the ordinate the original 1 cm is equivalent to the lg-p difference of ≈ 0.14, on the abscissa the original 1 cm is equiva-

---

[36] With the reversal of the cycle, the generator becomes a high-pressure *absorber*, the absorber a low-pressure *generator*, the condenser a high-pressure *evaporator*, and the evaporator a low-pressure *condenser*. In addition, regulating valves become pumps and pumps become regulating valves.

One example for a heat pump application for the domestic heating with utilization of geothermal heat is outlined in Section 9.1[155] by optimizing the simultaneous harnessing of the potential of outdoor cold.[155]

lent to the $1/T$ difference of $\approx 1.56 \cdot 10^{-4}\,K^{-1}$ so that the quotient required for the determination of the evaporation heat results from the tangent of the curve ascent as follows:

$$\frac{\Delta\,lg\,p}{\Delta(1/T)} = \frac{0.14}{1.56\ 10^{-4}} \cdot \tan\alpha = 892\ \tan\alpha$$

Thereby, the evaporation (plus solution) heat r is obtained according to Equation 5.3 by inserting the molecular weight for ammonia-water vapor with the vapor composition ξ_{vapor} (mass fractions of ammonia) according to the formula

$$\mu = \xi_{vapor}\ 17 + (1 - \xi_{vapor})\ 18 \tag{Eq. 5.5}$$

For pure ammonia vapor ($\xi_{vapor} \approx 1^{37}$) thus results

$$r \approx 1.35\ 10^{3} \cdot tg\,\alpha\quad [kJ \cdot kg^{-1}]$$

5.4 Historical Prejudices Toward the Absorption Technology

As Altenkirch commented[5], C. v. Linde initially made it plausible[45] 'that the refrigeration capacity of the absorption machine had to remain smaller than the supplied amount of heat.' Later, Hans Lorenz[172] calculated the processes in the absorption machine and quantified the heat ratio (ζ = quotient of the utilizable cold Q_o to the supplied thermal heat Q_h) up to maximal 0.52...0.55. When this theoretical statement was outpaced during practical operation by the Habermann machine[171] of the company "Norddeutsche Eiswerke", R. Plank[98] finally evaluated the theoretical heat ratio of the 'ideal' absorption machine under the operating conditions of the Habermann machine with $\zeta = 0.625$. For that, he used the results of the measurements for the vapor pressure and solution heats of the ammonia-water solutions[81] which had been publicized meanwhile.

On the other hand Altenkirch's work showed that according to the First Law of Thermodynamics it results from the equality of the overall amount of heat supplied to the evaporator and generator and discharged from the condenser and absorber, respectively, that means from

$$Q_o + Q_G = Q_C + Q_A = Q \tag{Eq. 5.6}$$

the following: Considering Q as discharged from the absorber and condenser at the mean temperature $T = (T_C + T_A)/2$, thus follows from the entropy balance in the theoretically best, that means, reversible case

$$\frac{Q}{T} = \frac{Q_o}{T_o} + \frac{Q_G}{T_G} \tag{Eq. 5.7}$$

and, therefrom derivable an ideal value ζ for the heat ratio by using for Θ the reciprocal absolute temperature $1/T$

---

[37] valid up to an ammonia content of $\xi_{fluid} \approx 0.5$ in the liquid

$$\zeta = \left(\frac{Q_o}{Q_G}\right) = \frac{\dfrac{1}{T_o} - \dfrac{1}{T}}{\dfrac{1}{T} - \dfrac{1}{T_G}} = \frac{\Theta - \Theta_o}{\Theta_G - \Theta} \text{ or, also } = \frac{T_G - T}{T - T_o}\cdot\frac{T_o}{T_G} \quad \text{(Eqs. 5.8 and 5.9)}$$

with Q_G, Q_o: the absorbed thermal heat, and the absorbed 'cold'

Q: the sum of the emitted heats (cooling) from the absorber and condenser (resorber).

When the temperature differences $T_G - T$ and $T - T_o$ become smaller, the concentrations in the generator and condenser, and the pressures in the generator and absorber approach to each other, especially with the resorption machine (see Figure 5.8 later below); at the same time—due to the continuity of the solution plot— the heat ratio ζ according to Equation 5.8 approaches the value 1.

Thus, the value "1" for the heat ratio is indeed significant as it is independent from the pair of substance. As inferred from Equation 5.8, ζ could very well attain values greater than 1 when the temperature stroke $T_G - T$ is just larger than the temperature lift $T - T_o$.

Recognizing this interrelationship, Altenkirch sought for a means to break through the boundaries imposed by the solution plot of the absorption machine for these temperature ratios and the boundaries resulting according to Equations 5.8 and 5.9 for the heat ratio. He finally succeeded by introducing the internal heat transfer between the sorption vessels as well as by using the multistage principle (see below: section "Heat Transformations").

A simplified, rather instructive equation for the heat ratio by considering all four temperatures, that is in addition the condenser temperature T_C and the absorber temperature T_A, is obtained when the approximate constancy of the sorption heats with constant working medium concentration, which underlies the approximate straightness of the vapor pressure curves, is introduced by $Q_V = Q_o$ and $Q_A = Q_G$ into Equations 5.6 and 5.7—so the following formulae are obtained

$$\frac{Q_h}{T_A} + \frac{Q_o}{T_C} = \frac{Q_o}{T_o} + \frac{Q_h}{T_h} \Leftrightarrow Q_h\cdot\left(\frac{1}{T_A} - \frac{1}{T_h}\right) = Q_o\cdot\left(\frac{1}{T_o} - \frac{1}{T_C}\right) \quad \text{(Eqs. 5.10, 5.11)}$$

$$\zeta = \frac{Q_o}{Q_h} = \frac{\Theta_A - \Theta_G}{\Theta_o - \Theta_V} = \frac{T_G - T_A}{T_V - T_o}\frac{T_o\,T_V}{T_G\,T_A}{}^{38} \quad \text{(Eq. 5.12)}$$

$T_G - T_A$ is nowadays known as the driving temperature difference, and $T_C - T_o$ the temperature lift. At first, Altenkirch was occupied with the reduction of irreversibilities of the heat transformation processes for the one-stage absorption machine according to the schematic of Figure 5.2, both with external and internal heat transfer processes.

---

[38] More exactly, the reciprocal thermodynamic mean temperature Θ_m has to be inserted into Equation 5.12. For the absorber for example $\Theta_{Am} = (\Theta_{A,inlet} + \Theta_{A,outlet})/2$ is valid.

5.5   External Reversibility

5.5.1   Adaptability to sliding heat-carrier temperatures

In emphasizing the advantages of the absorption machine in contrast to the compression refrigeration machine, Altenkirch referred to the widening of temperature bands of sorption processes due to the reduction of the solution cycle whereby they are well appropriate for the reversible warming-up or cooling-down of external heat carriers like the brine and cooling medium. He recognized that compared with that the external heat transfers with the compression refrigeration machine—because they are following the Carnot process and emit or absorb heat only at temperature points (that means to condenser and evaporator)—must proceed in general with energy losses. To this end, the following:

Let be $T_{CW,inlet}$ the initial and $T_{CW,outlet}$ the final temperature of the cooling water, $T_{BR,inlet}$ and $T_{BR,outlet}$ the respective initial and final temperatures of the brine, then the heat transfer in the compression machine can only be taking place when the condenser temperature lies above $T_{CW,outlet}$ and the evaporator temperature beneath $T_{BR,outlet}$. The temperature lift $T_C - T_o$ which has to be brought up by the compression machine must be therefore greater than $T_{BR,inlet} - T_{BR,outlet}$.

When, contrary to that, the evaporating and liquefying processes each would occur in temperature intervals—which, in the best case, would be congruent with the cooling-down and warming-up intervals of the brine and the cooling water—so only a temperature lift of $(T_{CW,inlet}+T_{CW,outlet})/2 - (T_{BR,inlet}+T_{BR,outlet})/2$ would be necessary. The lift to be achieved thus would be decreased by $\Delta T_{CW} + \Delta T_{BR}$ whereby $\Delta T_{CW} = T_{CW,outlet} - T_{CW,inlet}$ and $\Delta T_{BR} = T_{BR,inlet} - T_{BR,outlet}$ be the heating-up and cooling-down intervals of the cooling water and the brine. Thus, the theoretical expense of work would be decreased by the factor

$$0.5 \cdot \frac{\Delta T_{CW}+\Delta T_{BR}}{T_C - T_o} \qquad \text{(Eq. 5.13)}$$

When a corresponding advantage with the compression machine should be achieved, the evaporator had to be replaced by a degasser and the condenser by a resorber and a solution cycle conducted through each of them [39].

In the year 1895[49] Hans Lorenz had already suggested another option of implementation. According to that, the CARNOT process of the compression machine should be replaced by the polytropic process named for him of Figure 5.3. The ideally achieved reduction of the expense of work is illustrated graphically: The cooling water, in both the Carnot and LORENZ process in the borderline case, is heated from T_2 $(=T_{CW,inlet})$ to T_3 $(=T_{CW,outlet})$ and the cooling brine

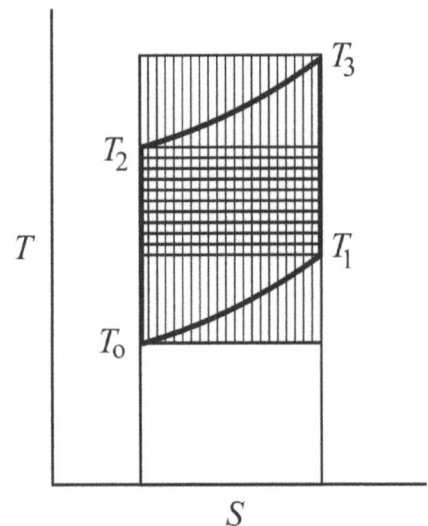

Fig. 5.3: Lorenz process

---

[39] See also Section 8 and Section 9, item 9.3.1.[37]

cooled from T_1 $(=T_{BR,inlet})$ to T_2 $(=T_{BR,outlet})$. The bolt solid lines mark the Lorenz cyclic process. In this borderline case of a reversible heating of cooling water (i. e. when the driving temperature difference for the heat transfer disappears), the differential heat quantity $C \cdot dT_x$, absorbed from the cooling water at a temperature T_x, matches exactly with the from the Lorenz process on the piece of curve $T_2 \rightarrow T_3$ at the same temperature to the cooling water emitted heat $T_x \cdot dS$. Thus follows for the entropy change

$$C \cdot dT_x = T_x \cdot dS \quad \rightarrow \quad \frac{\partial}{\partial T_x} S = \frac{C}{T_x} \qquad \text{(Eq. 5.14)}$$

When the same is assumed for the change of state of the cooling brine from T_1 to T_0 so the work to be expended sinks onto the bolt-framed surface which is always smaller than the vertical hatched surface of the Carnot process. Quantitatively, the *relative* saving on work—after inserting the thermodynamic mean temperatures—is more precisely constituted by

$$\frac{W_{Lorenz}}{W_{Carnot}} = \frac{T_0}{T_{om}} \cdot \frac{T_m - T_{om}}{T_3 - T_0} \approx \frac{T_m - T_{om}}{T_3 - T_0} \qquad \text{(Eq. 5.15)}$$

with: T_{om} : thermodynamic mean temperature of the curve piece $T_1 \rightarrow T_0$
T_m : thermodynamic mean temperature of the curve piece $T_2 \rightarrow T_3$

Regarding the viability of this process, Altenkirch referred to an idea of Hans Lorenz who remarked, with view to common applications of Carnot processes in *steam engines and compression refrigeration machines*, that a closed series of Carnot processes would come close to the ideal polytropic process. This way, the *steam engine* had tried to use successfully the application of manifold expansions with intermediate superheating. Also with *compression refrigeration machines* separate compression stages were applied, inserted between different temperatures. The further development of the machines along these lines failed however in the case of more than two stages[40] due to the increasing complexity of the machinery and rise in cost of the facilities.

5.5.2 Heat Transformation in Sorption Processes (Lorenz Process)

Practically, the heat flow per degree of the sorbing solution is in general erratic, that means, distinctly non-temperature-dependent as shown in Figure 5.6 whereas the absorption of heat of a cooling/heating medium like brine or water remains per degree almost constant. As a result, the virtual heat capacity of the boiling solution with its heat transfer to the brine becomes temperature-dependent. The complete heat transfer evaded therefore immediate computability through closed mathematical expressions. Altenkirch therefore calculated the processes of cooling/heating by external media and the reachable final temperatures of the media with iterative arithmetic techniques[44].

This shall be shown by the binary system ammonia/water in Table 5.1 (data according to Figure 5.2) using Altenkirch's chosen example of an absorber with an initial tem-

---

[40] with more than two stages (Lotz' personal comment)

perature of absorption at 70 °C and a final temperature at 20 °C with constant pressure at p_o = 2 ata (with an evaporator temperature at t_V = – 19 °C).

In this example, the sorbing solution enters at the right end of the absorber (in Table 5.1 beginning from the right hand side) into the i = 10th stage with $T_{BR,inlet}$ (stage 10) = 70 °C (on the right, outside of the table). Up to a temperature $T_{BR,outlet}$ (10) = 65° C[41] the heat quantity ΔQ (10) = 11.2 kcal/kg H_2O must be discharged from the solution, analogously for the other stages. The cooling water, streaming in counter-current to the solution, should enter according to the boundary conditions with $T_{CW,inlet}$ = 15 °C (at the cold end, 2nd column, 4th line) into the absorber and leave with $T_{CW,outlet}$ = 60 °C. By uptake of 11.2 kcal (2nd line, 10th column) the cooling water is warmed up at the last stage from $T_{CW,inlet}$ (10) = 56.9 to $T_{CW,outlet}$ (10) = $T_{CW,outlet}$ = 60 °C, analogously at the preceding stages.

Tab. 5.1: Stage calculation of the temperature course for the cooling water and the absorbing solution

Column i	1	2	3	4	5	6	7	8	9	10	11
$T_{BR,outlet}$ (i) [°C]	20	25	30	35	40	45	50	55	60	65	Line
$\Delta Q(i)$		29.9	25.8	22.4	19.8	17.6	15.9	14.3	13.1	12	11.2	1
$Q(i)/Q_{whole}$ [%]		16.4	30.5	42.8	53.7	63.4	72.1	79.3	86.5	93.1	100	2
$T_{CW,inlet}(i)$		15	22.4	28.7	34.3	39.2	43.5	47.4	50.7	53.9	56.9	3
$T_{BR,out}(i)$-$T_{CW,in}(i)$		5	2.6	1.3	0.7	0.8	1.5	2.6	4.3	6.1	8.1	4
$T_{BR,inlet}(i)$-$T_{CW,outlet}(i)$		2.6	1.3	0.7	0.8	1.5	2.6	4.3	6.1	8.1	10	5
$\tau_m(i)$		3.68	1.82	0.92	0.71	1.05	1.89	3.15	4.77	6.65	8.84	6
$\Delta Q(i)/\tau_m(i)$		8.1	14.2	24.3	28	16.7	8.4	4.5	2.8	1.8	1.3	7
$(\Sigma \Delta Q)/\tau_m(i)$		8.1	22.3	41.5	74.6	91.3	99.7	104	107	109	1	8

In general, the warming-up of the cooling water with the ith stage is determined by its entry temperature (specific heat of water, set = 1)

$$T_{CW,outlet}(i) = T_{CW,inlet} + (T_{CW,outlet} - T_{CW,inlet}) \cdot \frac{Q(i)}{Q_{whole}} \qquad \text{(Eq. 5.16)}$$

Herein, $Q(i)$ indicates the integral heat emission of the solution divided by the heat absorption of the cooling water, calculated from the initial part of the absorber until the ith stage. Then, $Q_{whole} = Q(10)$ is the heat quantity total emitted from the solution in the absorber. The share $Q(i)/Q_{whole}$ of the absorption total of heat for the cooling water up to the stage i is indicated in line 2. According to Equation 5.16 therefrom the values in the lines 4 and 5 result one by one.

For the calculation of the cooling surface, the mean temperature difference $\tau_m(i)$ between solution and cooling water in the several stages is determined approximately according to the following formula which is decisive for the heat transfer

---

[41] value entered in the i = 10th column of the 2nd line

$$\tau_m = \frac{(T_{BR,outlet}(i) - T_{CW,inlet}(i)) - (T_{BR,inlet}(i) - T_{CW,outlet}(i))}{\log \dfrac{T_{BR,outlet}(i) - T_{CW,inlet}(i)}{T_{BR,inlet}(i) - T_{CW,outlet}(i)}} = \frac{T_{CW,outlet}(i) - T_{CW,inlet}(i) - 5}{\log \dfrac{T_{BR,outlet}(i) - T_{CW,inlet}(i)}{T_{BR,inlet}(i) - T_{CW,outlet}(i)}} \quad \text{(Eq. 5.17)}$$

wherein the parenthesized differences in the first quotient mean in general the corresponding temperature differences between the solution and cooling water at the hot or cold end of a stage, respectively. In the second quotient the constant warming-up of the brine by $T_{BR,inlet}(i) - T_{BR,outlet}(i) = 5\ °C$ which underlies the stage calculation (size of stages) is explicitly considered.

For the predefined parameters $T_{CW,inlet}$ and $T_{CW,outlet}$ of the stage calculation negative differences $\tau_m(i)$ in the absorber can be possible, which shows the inadequacy of the corresponding parameter selection.

The sequence of the τ_m values (line 7) together with the corresponding partial amounts of absolute values of heat ΔQ which will be transferred allow statements about in the sum and in the corresponding stage required heat transfer capacity $(\Delta Q / \tau_m)$. Smaller τ_m values in the absorber interior can lead mathematically to very large subarea shares of the cooling system of the absorber so that the inlet temperature of the cooling water must be decreased and/or the cooling water quantity enlarged.

The necessary heat transfer capacity in stage i is shown in line 8; of which summation $Q_{whole} = \Sigma(\Delta Q / \tau_m)$ in line 9 is the needed heat transfer capacity total. From that, the mean value $\bar{\tau}_m$ of all temperature differences τ_m can be calculated averaged over all segments of the absorber

$$\frac{Q_{whole}}{\bar{\tau}_m} = \sum \frac{\Delta Q}{\tau_m} \quad \text{(Eq. 5.18)}$$

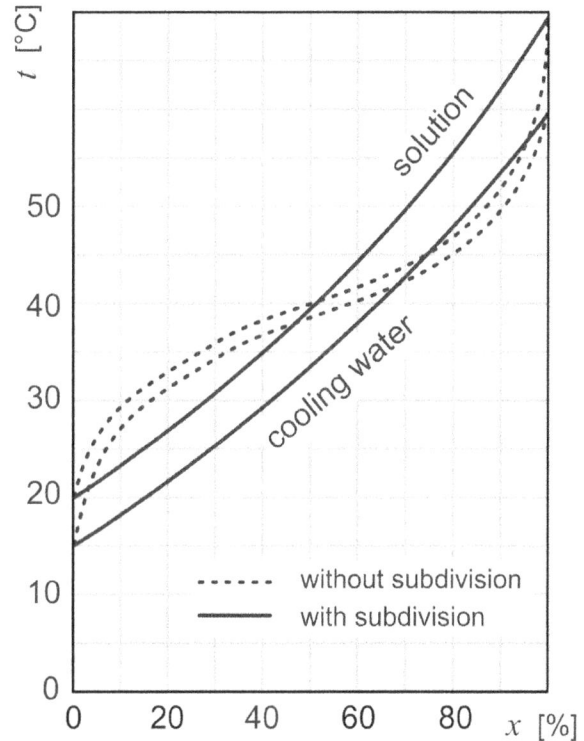

Fig. 5.4: Temperature course dependent on the position on a cooling surface assumed to be linearly extended

This mean or its reciprocal $1/\bar{\tau}_m$, respectively, is for the given temperatures $T_{CW,inlet}$, $T_{CW,outlet}$ a measure for the requirement of cooling surfaces (line 8). In the exemplary case, $\bar{\tau}_m$ amounts to 1.65 °C with considerably larger values at the ends. That illustrates the non-conformity of the warming-up process of the cooling medium with the actual heat turnover of the solution per degree according to Figure 5.6. Therefore, the mean temperature difference cannot be arbitrarily diminished due to the enlargement of the heat transfer capacity so that a residual irreversibility of the heat transfer is unavoidable (see Figure 5.4).

5.5.3 Subdivision of the solution cycle

Altenkirch demonstrated that this disadvantage can be mitigated by dividing the flow-rate of the solution into segments of the absorber. He achieved this by subdivision of the solution cycle as shown in Figure 5.5. Therein the poor solution passes the rerouting in the generator, afterward the heat exchanger, the regulating valve, and enters then at the hot side of the absorber. Each of these three partially enriched heat flows leaves its absorber segment and passes a solution pump. Afterward, each flow is conducted back through the absorber to the hot end of it, further warmed up in the common heat exchanger, and is fed to the assigned segment of the generator.

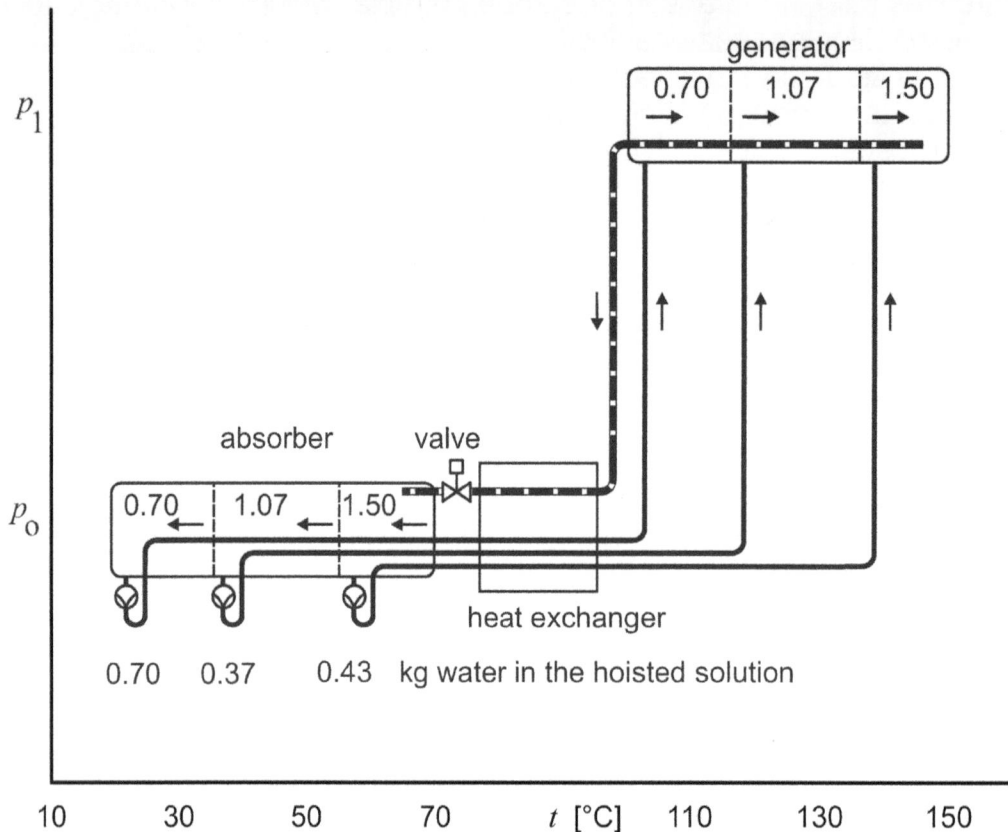

Fig. 5.5: Subdivision of the solution cycle

The absorber is divided into three temperature zones, from 20 - 35, 35 - 55, and 55 - 70 °C. The solution quantities circulating in these 3 intervals are so dimensioned that the absorber in these 3 temperature intervals converts equal amounts of heat. Thereby, the overall converted heat quantity amounts as in the previous example to 128.1 kcal/kg$_{solution}$, likewise the in- and outlet temperatures of the cooling water are equal.

The resultant temperature course of the absorber solution is outlined with the upper solid-line curve in Figure 5.4, that one of the cooling water with the lower curve.

To compare, the temperature courses without subdivision of the solution cycle are outlined with the broken-line curves. Due to the subdivision of the solution cycle the mean temperature difference rises from 1.67 °C up to $\bar{\tau}_m = 6.7$ °C, accordingly the requirement for the cooling surfaces decreases and, conversely, due to the reduction of the mean temperature difference with corresponding increase of the heat transfer surface the irreversible losses can further be reduced, respectively, that means the amount of cooling water in terms of approximation of reversibility with the Lorenz process can be reduced due to the reduction of the temperature differences at the beginning and end of the absorber. Respective statements are valid for the generator.

The preceding considerations demonstrate that for a predefined process to conduct an optimization of the cost total per hour, consisting of the operating costs for the cooling water, and the interest rate and amortization of the facility, by taking into account the degree of utilization, can be useful (see[44], page 81 et sequ.)

5.6 Enlargement of the Internal Reversibility "Endoreversibility"
Preheating/-Cooling of Solution and Working Medium Stream[160]

5.6.1 Removal of Losses in the Solution Heat Exchanger

Also with ideal heat transfer of the solution heat exchanger, both due to the different mass streams of the poor and rich solution and due to the sliding temperatures in the sorption vessels, losses occur due to intermixture of the supercooled or superheated solutions with the boiling solution. Altenkirch's countermeasures should firstly be represented by quoting the main claim of his patent from August, 1911[160], in his preferred concise form of writing,

'Absorption refrigeration machine, ..., characterized thereby that the hot solution after its degasification is rerouted first in counter-current through the generator, and the cold solution first in counter-current through the absorber, after either of them leave the generator or absorber.'

Solution Rerouting in the Generator

With his *rerouting of the poor solution through the generator Altenkirch* achieved the pre-cooling of the poor solution in counter-current and in thermal contact with the desorbing (out-gassing) solution in the generator already before the heat exchanger.

This rerouting starts at the hot end of the generator and finishes at its cold one. The rerouting regains that part of the amount of thermal heat which alone is needed for the increase of temperature of the out-gassing solution due to its rising boiling temperature. Overall, the poor solution which leaves the generator reaches as a result its approximate inlet temperature again and has thus the effect of an additional *heating* for the generator—simultaneously a significant relief is provided to the heat exchanger at its hot side and the rectification (counterflow distillation) of the generator is intensified.

The curves which run to the upper edge of the diagram in Figure 5.6 describe the heat quantities per degree and per kg water, released or bound during the sorption processes in the solution. The reference to the water content is possible since the vapor—apart from its immediate vicinity to the saturation line of pure water—virtually only contains ammonia (working medium) so that the water content in the solution barely changes—also the rectification of the vapor in the approximate equilibrium with the solution—when it is enabled or intensified by constructive measures (sheets of metal)—contributes to it.

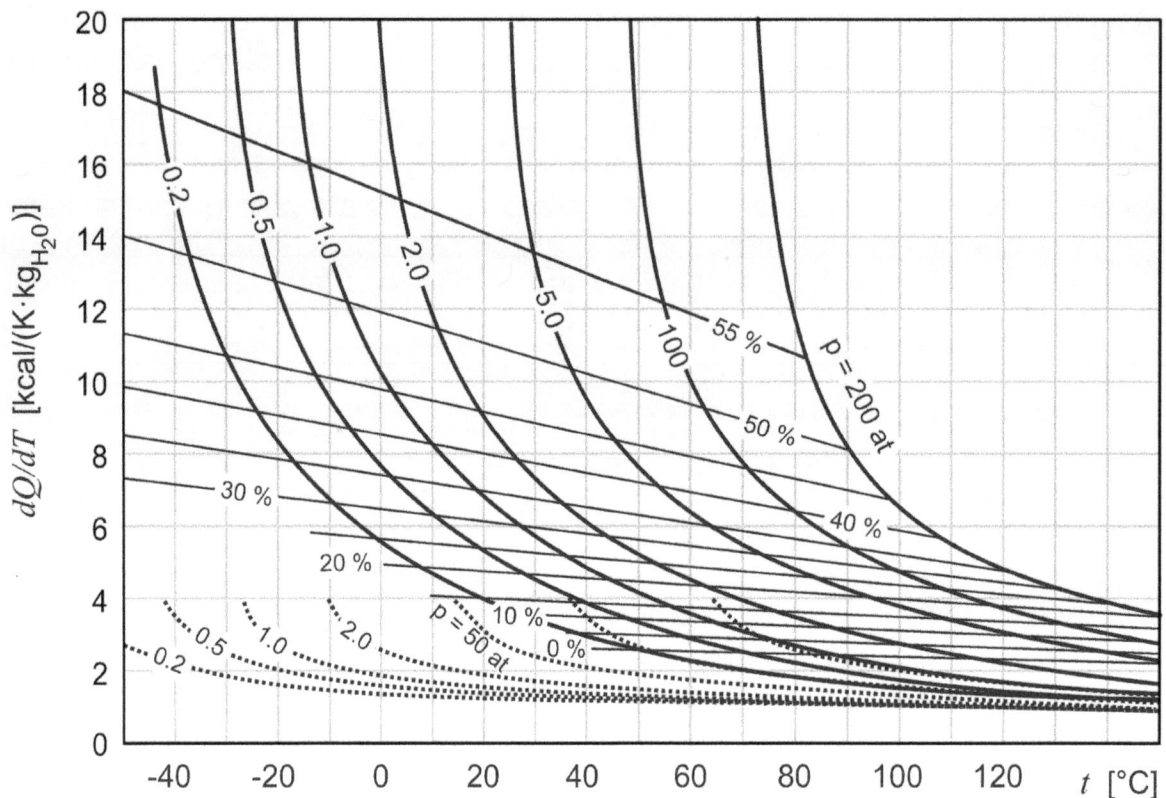

Fig. 5.6: The change of the enthalpy per Kelvin for an aqueous ammonia solution

The lower curves illustrate the heat capacity of the solution with a water content of 1 kg and a certain working medium content which is equivalent to the equilibrium (boiling) state with the entered pressures and plotted temperatures on the abscissa. These heat capacities practically do not change either with the super-heating and super-cooling so that also the absolute values of heat for the warming-up and cooling-down can be taken from the remaining reroutings (see further below) and heat exchangers.

Solution Rerouting in the Absorber

Similarly, after finalizing its sorption process in the absorber and after pressure boosting in the solution pump from p_o to p, the rich solution, shortly after its discharge from the cold side of the absorber, can be rerouted through the absorber by preheating in counter-current heat exchange to the absorbing solution for its entry into the generator. Thus, there is a saving on cooling medium for the absorber. The rich solution, coming from the rerouting through the absorber, reaches in the ideal case the initial temperature $T_{A,inlet}$ of the poor solution which enters from the warm side of the absorber at pressure p_o so that significant relief can be provided to the heat exchanger also on its cold side. Since the heat capacity of the rich solution is larger than that of the poor solution, a part of the sorption heat of the absorber for the warming up inside the absorber is needed—in contrast to the rerouting of the poor solution in the generator—whereby—as it is analogously in the generator—also here the solution stream takes over a part of the function of the cooling medium.

Calculations of the Absolute Values of Heat to be Transferred in the Reroutings

The previously discussed elimination of the internal irreversibilities in the solution cycle due to counter-current heat exchangers has two effects:

- The temperatures at both the cold and hot end of absorber and generator approach the respective equilibrium temperatures of the poor and rich solution at the pressures p and p_o due to the reroutings.
- As a result, the cooling medium can be more intensively heated in the absorber, the heating medium more intensively cooled in the generator. Both contributes to the improved utilization of the media and the saving on the need for these media, respectively.

These interrelationships shall be explained in the following example restricted to the *absorber*. In this example, two cases (I and II) of the working process with ammonia/water as the binary blend used with different widths of the temperature band inside the sorbing vessels (say: different size of the specific solution cycle f, it means, of the circulating amount of poor solution per kg working medium) will be considered and in both cases the theoretically possible saving on energy of $\dot{Q}_l$ and $\dot{Q}_h$ calculated.

The specific solution cycle will be determined due to the working medium concentrations ξ_r and ξ_p of the rich and poor solution according to the known formula

$$f = \frac{1 - \xi_p}{\xi_r - \xi_p}$$
(Eq. 5.19)

The pressure in the absorber is chosen with $p_o = 2$ bar as in the example of Figure 5.2. Therewith, the equilibrium temperatures at the beginning and end of the absorber and the width of ΔT of the sorption interval are fixed. The rerouting of the rich solution through the absorber per kg working medium yields, as a result, the absolute value of the caloric amount of cooling

$$\dot{Q} = f \cdot \Delta T \cdot c$$
(Eq. 5.20)

(when c specifies the specific heat of the rich solution). For comparison, one is referred to the converted heat flow total $\dot{Q}_A$ in the absorber. Through integration of the respective curve in Figure 5.6 in the temperature interval ΔT with the chosen pressure p_o, it approximately results in

$$\dot{Q}_A \approx m_{H_2O} \cdot \Delta T \cdot \frac{\left. \dfrac{d\dot{Q}}{dT} \right|_{\xi_r, p=2} + \left. \dfrac{d\dot{Q}}{dT} \right|_{\xi_p, p=2}}{2}$$
(Eq. 5.21)

whereby in Figure 5.6 the multiplication defines the reference to 1 kg water

$$m_{H_2O} = f \cdot (1 - \xi_r)$$
(Eq. 5.22)

The results are compiled in Table 5.2. The last line specifies still the shares of the converted heat as a percentage.

Removal of the Heat Exchanger

Altenkirch documented this effect in claim 2 of his Patent[160], 'Absorption refrigeration machine, ..., characterized thereby that in order to avoid the heat exchanger the degasification in the generator has to be continued so far that the highest absorber temperature approximates to the lowest generator temperature.'

Thereby, the occurrence of the rerouting of the solution in the driving part is a prerequisite. Altenkirch wrote in this context, 'The stated advantages of the rerouting became the more distinguished the greater the approximation to the Lorenz process due to the augmentation of the degasification width. Thereby, a point was reached that the heat exchanger could be omitted at all, and the heat transfer for the whole pre-heating and pre-cooling of the solutions could take place in the absorber and generator.' The schematic of this disposition shows Figure 5.7.

Tab. 5.2: Processes with rerouting of the solution

		Case 1	Case 2
$T_{A,inlet}$	[°C]	70	38
$T_{A,outlet}$	[°C]	20	20
ΔT	[K]	50	18
c	[kcal·kg^{-1}·K^{-1}]	1.1	1.1
ξ_r	[-]	0.45	0.45
ξ_p	[-]	0.18	0.33
$(dQ/dT)_{p=2,\xi_r}$	[kcal·K^{-1}·kg$_{H2O}$$^{-1}$]	9.2	9.2
$(dQ/dT)_{p=2,\xi_p}$	[kcal·K^{-1}·kg$_{H2O}$$^{-1}$]	3.9	6.4
$Q_{po=2\,ata}$	[kcal·kg$_{NH3}$$^{-1}$]	547	431.1
f	[-]	3.04	5.58
q	[kcal·kg$_{NH3}$$^{-1}$]	167.0	110.6
m_{H2O}/m_{NH3}	[-]	1.67	3.07
q/Q	[%]	30.5	25.6

Solution Prerouting in Resorber/Degasser of a Resorption Machine

The introduction of the solution cycle through condenser and evaporator of a compression refrigeration machine[42] was already proposed by Osenbrück. This connection had never been executed, but Altenkirch recognized the advantages with combinations of absorption machines. For a convenient discrimination of an absorber which takes over the function of a condenser Altenkirch introduced the term "resorber" and for a generator which takes over the function of an evaporator the term "degasser." The whole assembly was henceforth called *resorption machine* (Figure 5.8).

---

[42] Patent D.R.P. 84084

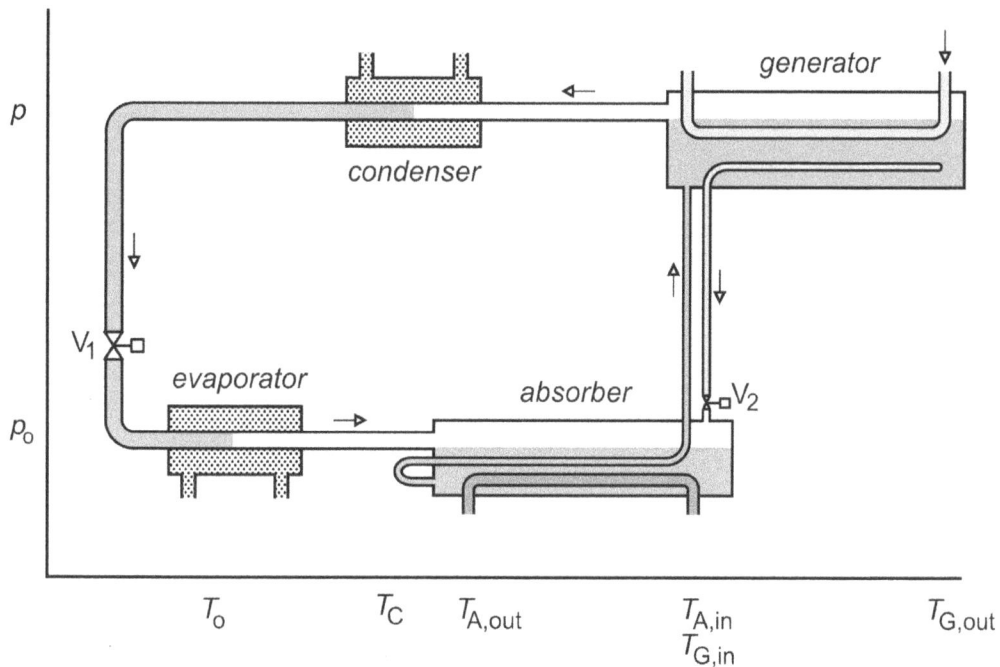

Fig. 5.7: Removal of the heat exchanger

Since the solution cycle in the cooling part proceeds inversely to that of the driving part, rerouting becomes prerouting—otherwise the effects remain the same. The working diagram of the resorption machine is outlined in Figure 5.9.

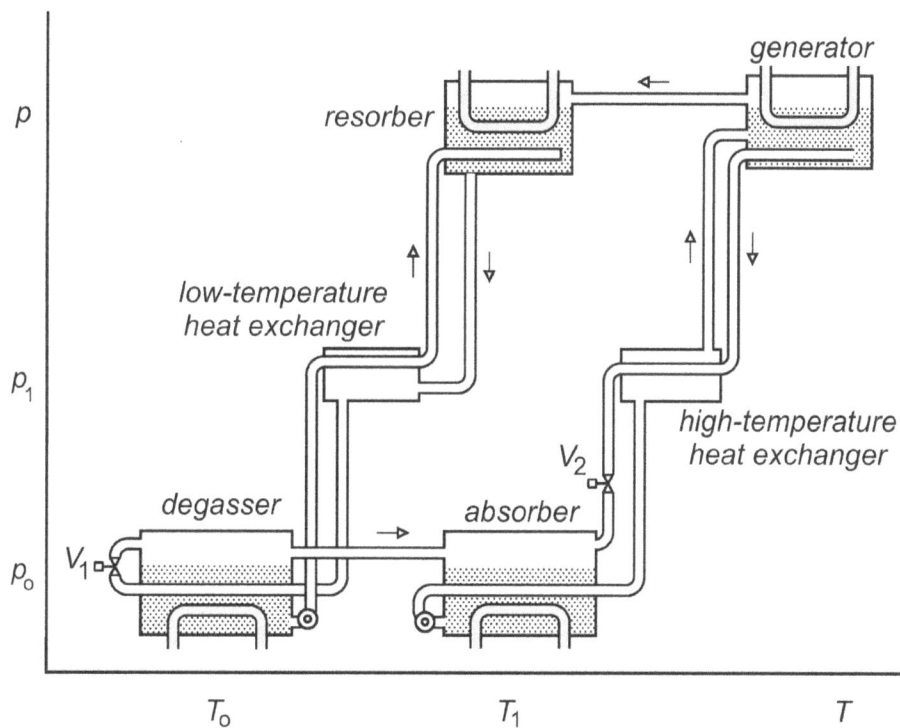

Fig. 5.8: Schematic of a resorption machine with pre- and rerouting of the solution

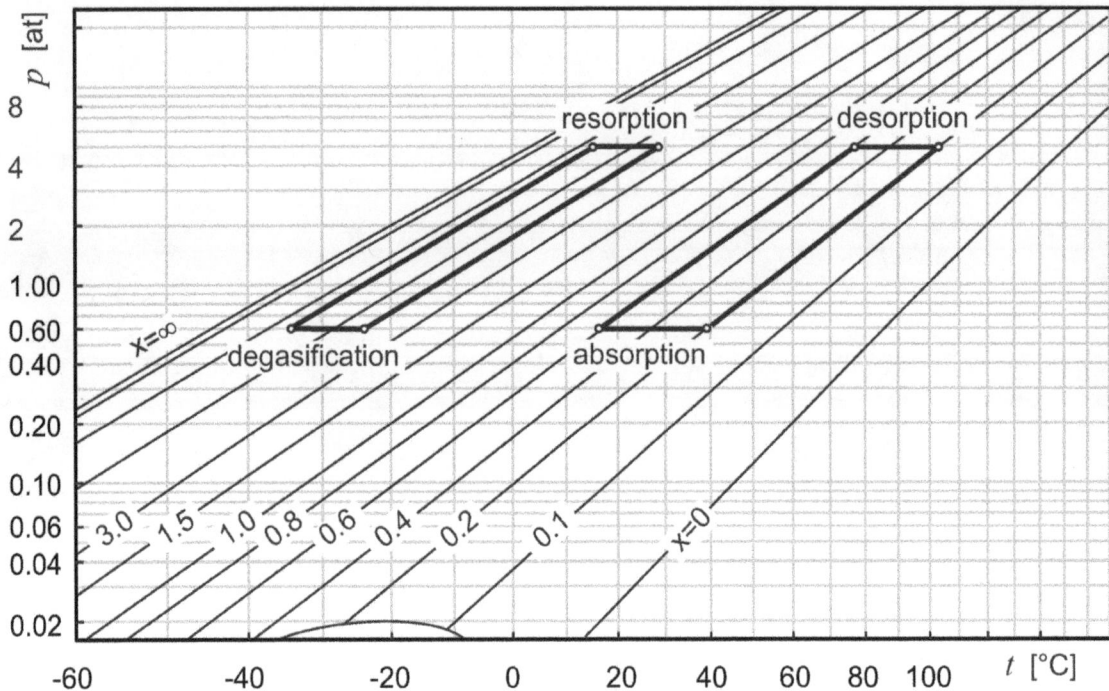

Fig. 5.9: Working diagram of the resorption machine

5.6.2 Cycle of the Working Medium in the Absorption Machine

Between the vapor which flows from the generator to the condenser and the working medium which is contained as a component in the stream of the rich solution from the absorber to the generator according to Altenkirch a counter-current heat exchanger also should be inserted.

This *steam cooling by preheating of one part of the rich solution coming from the absorber* implies (in terms of Altenkirch) that the preheating of the other part of the rich solution is taken over by the poor solution leaving the cold side of the generator after rerouting. This measure yields a further approximation for the reversibility.

For implementation of such a heat transfer actually three flows of heat have to be brought into contact (according to the first author of this book):
· the rich solution from the absorber
· in counter-current to it the poor solution from the generator, and
· in direct-current to the latter the generator vapor (at the same time the rectification, i. e. the return flow of the solvent which is thawed out from the vapor, must be allowed[161]).

Compared to this, a somewhat simplified type of generator with a rectifier and vapor cooler by Niebergall is shown in Figure 5.10. The main parts of this device are the *steam condenser, separation column, generator,* and *collecting pot.* The vapor enters from underneath the filling layers of the *separation column,* and later is conveyed into the upper part equipped with bubble trays. Special guiding assemblies cause the rich solution entering the middle of the apparatus to be distributed evenly onto the inner surface of the tubes which are around, which the flow passes with the heating steam.

A condensate precooler (not outlined in the picture) should be, according to a proposal of Altenkirch, inserted between the condensate which leaves the liquefier and the vapor from the evaporator. Due to the never quite complete rectification a residual water content of the condensate can be expected so that the unevaporated liquid remainder of the condensate together with the gas flow from the evaporator cause a pre-cooling of the condensate in the condensate precooler.

This residual solution provides, as Altenkirch emphasizes, an especially effective contribution to the pre-cooling of the condenser. A temperature measurement in this solution stream can be used as an indicator for the degree of rectification according to Altenkirch[44] and can serve for the control of the cooling capacity of the steam condenser assembled in the course of the working medium flow from the generator (Figure 5.10). The working medium and the co-evaporating solution return in the same way from the condensate precooler/reboiler to the absorber (not shown).

5.6.3 Other Cycles

When—as described later—an inert gas cycle is intended without using a mechanical pump and an expansion valve for pressure equalization, the cycle will proceed through the absorber and evaporator (for example in case of the *diffusion machine*—see Section 5.9: "Response"). Also small quantities of inert gas can be conveyed due to absorption into the rich solution and are released in the generator again. In the long run, these quantities of gas would obstruct the condenser; thus they have

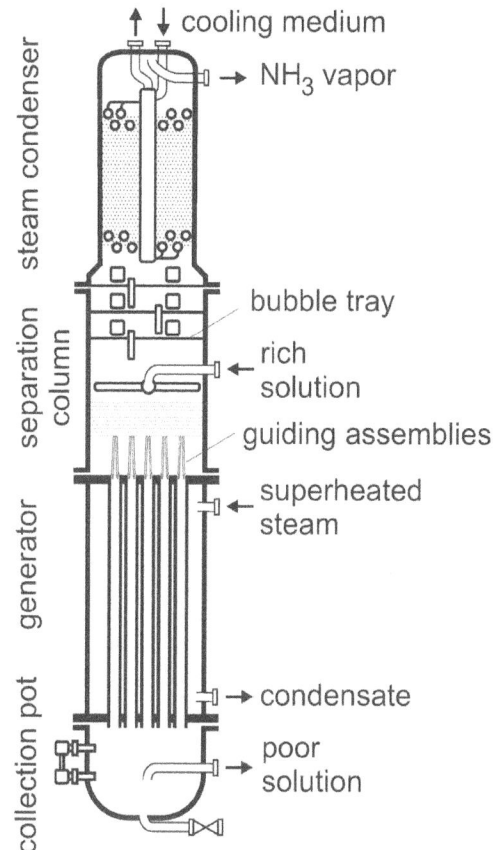

Fig. 5.10: Vertically arranged steam-heated ammonia generator

to be conducted back from there into the evaporator/absorber cycle of the inert gas due to washing out—for example with the poor solution (which will not be further discussed here). Into this inert gas cycle a *gas heat exchanger* is inserted. According to VON PLATEN[105] the cycle is powered by molecular weight differences between the rich and poor gas, or mechanically, for example by a nozzle[82].

Furthermore, there are dissociation products of the working medium or solvent, especially initiated due to catalytic effects of the construction materials (for more details see Section 8 and Section 9.5 "Further Development of the Sulfuric Acid/Water Chryothermal Apparatus"). The formation of these compounds must be either largely limited, or the products must be collected and occasionally removed out of the cycle, or converted into non-interfering compounds and then deposited.

As Altenkirch emphasized, the large scale deployment of absorption machines, 'almost failed due to this problem and was only enabled by meritorious investigations of E.C. McKelvy and Aaron Isaacs who developed a prophylactic (by adding 0.2 % sodium or potassium bichromate to generate a protective layer on the iron surface) that prevented the corrosion of iron.'

5.7   Heat transformations

The improvements, as far as mentioned here, introduced or proposed by Altenkirch which, in essence, refer to the removal of irreversibilities inside the circular process of the absorption machine and its coupling elements to external media, led him eventually as a result of their thermodynamic consequence to his spectacular connections which are capable of solving the reversible realization of predefined energy-industrial tasks subsumed under the term *heat transformation*.

In the following, Altenkirch's proposed single- and multistage connections of cycles of absorption machines are demonstrated. Here the term *reversibility* means the approximation to the heat ratio according to Equation 5.9 for the realization of concrete energy-industrial objectives of the use of available external cooling and heating media at given temperatures. The methods applied by Altenkirch go beyond the earlier mentioned re-/prerouting of the sorption solutions, and the rectification and preheating/-cooling of the fluid and vaporous working medium. There are measures such as:

. the enlargement of the degasification widths of the internal heat transfer, and
. the cascade design of the driving/cooling part by heat transfer between the stages of the cascade—as already mentioned above—with thermoelectric systems (see Section 3: "Thermocouple for Reversible Heating"), and reversible heating (see Section 4: "Reversible Heating").

5.7.1   Adjustments[43] due to Internal Heat Transfer

Claim 3 of his Patent[160] on August 12, 1911 had the wording, 'Absorption refrigeration machine, …, according to claim 1, characterized thereby that for the reduction of the heat expenditure of the same cooling capacity (or also working capacity) the degasification in the absorber is driven so far that the highest absorber temperatures exceed essentially beyond the lowest of the generator and that the coldest parts of the generator and the hottest parts of the absorber are in thermal contact with each other …'

That was Altenkirch's significant pioneering deed. Thus, it was possible to break through conventional borders of effectiveness. It was especially favorable that for the first time the generated refrigeration could be greater than the supplied thermal heat (heat ratio $\zeta > 1$), (see the beginning of Section 4.2).

Figure 5.12 shows the mechanism of realization of the 'overlapping of temperatures' in the driving part of an absorption machine. For the explanation of the thermal effect was provided, as quite often made as a simplified assumption, that the vapor pressure curves ran parallel within the solution plot between absorber and generator

---

[43] That means adjustments of the refrigeration machine to the thermodynamically best enabled utilization with given external thermal boundary conditions (utilizable temperatures of natural heat sources (waste heat from industrial plants, solar collectors, the ground water, etc.) for the achievement of technical purposes (cooling for preserving, freezing, and the like).

pressure, that is, within the whole concentration interval of the revolting solutions among the *absorbers, resorbers, degassers* and g*enerators.* According to CLAUSIUS-CLAPEYRON, the heat flows per degree are then the same magnitude in absorber and generator—a simplification whose error may be neglected compared to the comparatively strong caloric effects of the multistage designs:

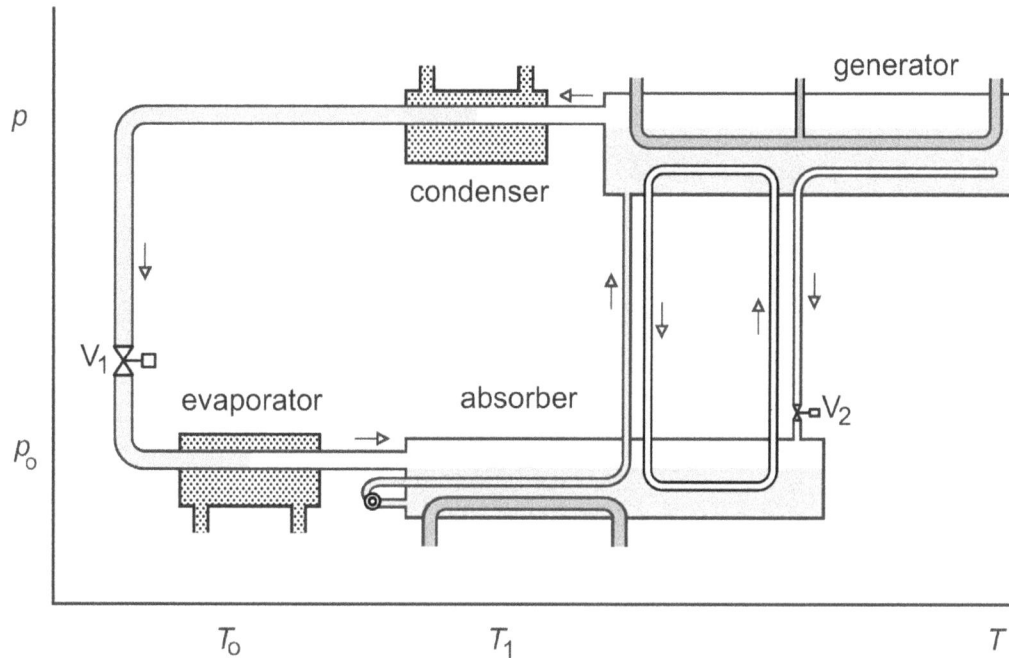

Fig. 5.11: Schematic 'Overlapping in the driving part'

As can be seen, the width total $\Delta\Theta_{width}$[44] of the absorber or generator, respectively, on the horizontal abscissa is the sum of the width $\Delta\Theta_{over}$ of the overlapping interval in absorber and generator and the temperature stroke $\Delta\Theta_{stroke}$—the difference of the Θ values between the begin of the generator and the end of the absorber, or the end of the generator and the begin of the absorber, respectively—or, (*due to the presupposed parallelism of the vapor pressure curves*) between the mean temperatures of the generator and absorber

$$\Delta\Theta_{width} = \Delta\Theta_{over} + \Delta\Theta_{stroke} \qquad \text{(Eq. 5.23)}$$

With this defined preconditions, the converted absolute values of heat in the absorber and generator are equal. They are denoted here with $\dot{Q}_{AG}$. Also the vapor pressure curve of the pure working medium in the solution plot is provided linearly, and thus also the absolute values of the evaporator and condenser heats are presupposed equal and denoted with $\dot{Q}_{VC}$.

----

[44] $\Theta = 1/T$ see Nomenclature

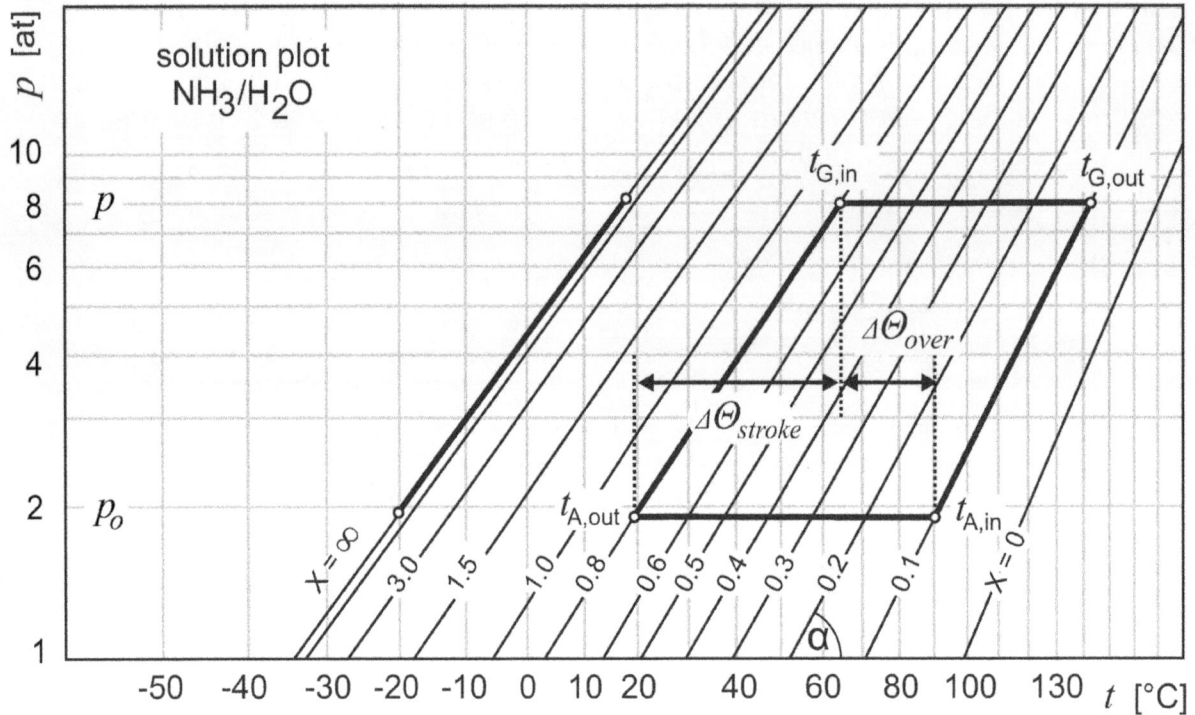

Fig 5.12: Working diagram 'Overlapping in the driving part'

In absorber and generator the respective shares $\Delta\dot{Q}_{over}$ and $\Delta\dot{Q}_{stroke}$, of the entire heat flow $\dot{Q}_{AG}$ are thus proportional to the lengths of the respective Θ intervals $\Delta\Theta_{over}$ and $\Delta\Theta_{width}$. When a "degree of overlapping" is introduced due to

$$\delta = \frac{\Delta\Theta_{over}}{\Delta\Theta_{stroke}} \text{ then is valid } \Delta\dot{Q}_{over} = \frac{\delta}{1+\delta} \cdot \dot{Q}_{AG} \qquad \text{(Eqs. 5.24, 5.25)}$$

For the heat ratio the supplied remainder of thermal heat to the generator is of interest

$$\Delta\dot{Q}_{over} = \dot{Q}_{AG} - \dot{Q}_{AG \text{ over}} = \frac{1}{1+\delta} \cdot \dot{Q}_{AG} \qquad \text{(Eq. 5.26)}$$

The heat ratio $Q_{VC}/Q_{AG \text{ over}}$ therefore is increased with overlapping by

$$\zeta_{over} = \zeta \cdot (1+\delta) > \zeta \qquad \text{(Eq. 5.27)}$$

The same result is obtained when the reciprocal of the absolute mean temperatures, with overlapping $\Theta_{Gm \text{ over}}$, $\Theta_{Am \text{ over}}$ and without overlapping Θ_{Gm}, Θ_{Am}, which is inferred from Figure 5.12, are inserted into Equation 5.23

$$\Theta_{\text{Gm over}} - \Theta_{\text{Am over}}$$

$$= \Theta_{\text{Gm}} + \frac{\Delta\Theta_{\text{over}}}{2} - \left(\Theta_{\text{Am}} = \frac{\Delta\Theta_{\text{over}}}{2} \right) \quad \text{and one obtains} \qquad \text{(Eq. 5.28)}$$

$$= \Delta\Theta_{\text{stroke}} + \Delta\Theta_{\text{over}}$$

$$\zeta_{\text{id over}} = \frac{\Delta\Theta_{\text{stroke}} + \Delta\Theta_{\text{over}}}{\Delta\Theta_{\text{lift}}} = \zeta_{\text{id}} \cdot (1 + \delta) \qquad \text{(Eq. 5.29)}$$

where $\Delta\Theta_{\text{lift}} = \Theta_{\text{V}} - \Theta_{\text{C}} =$ ("stroke" in the cooling part) $= \Delta\Theta_{\text{stroke}}$ is valid due to the set assumptions.

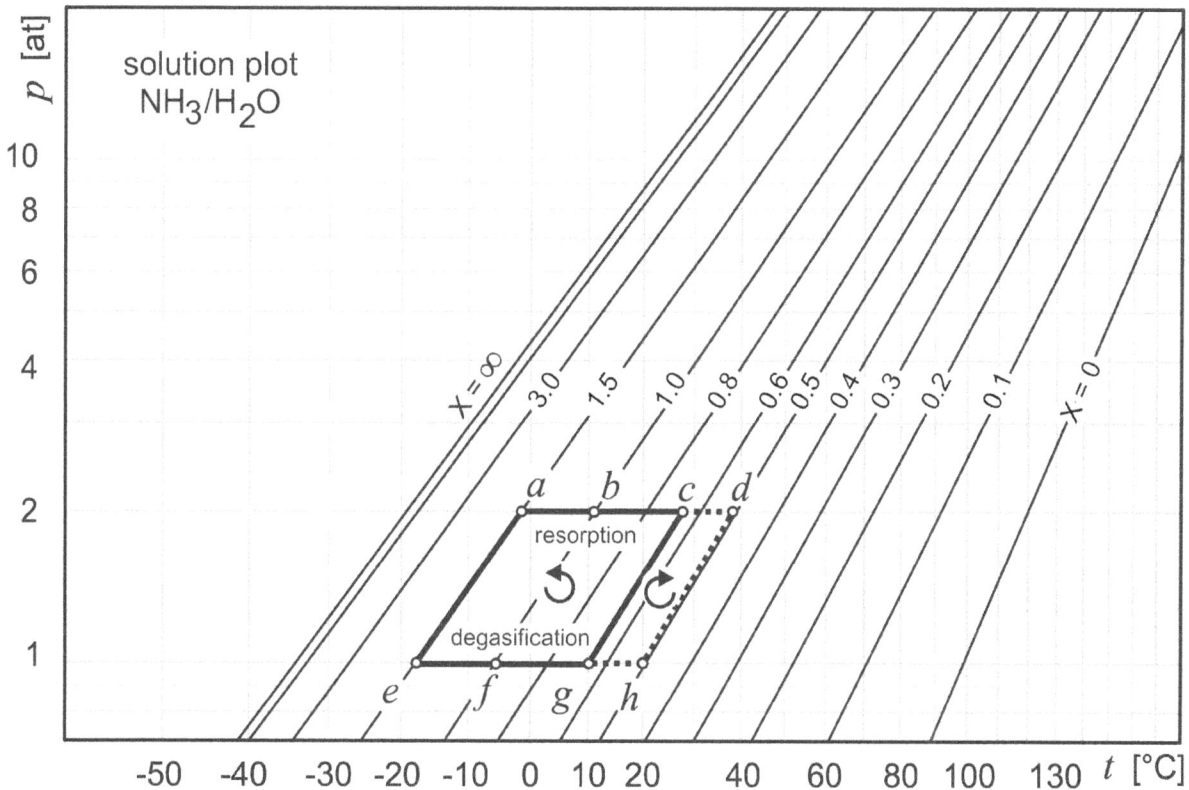

Fig. 5.13: Working diagram of the resorption machine with overlapping in the cooling part

In terms of the internal heat transfer between the hot part of the absorber and the "cold" part of the generator the discussed difficulty is similar to that experienced with the heat transfer from the absorber to the cooling medium (see Figure 5.12 and Table 5.1). The results of an iterative calculation of this heat exchange are shown in Figure 5.14 and the resulting heat ratio ζ is outlined for all discussed measures:
· The sole heat exchanger (lower solid-line curve for $p = 8$ and 10—see Figure 5.14). With growing degasification width the final generator temperature and therefore the load of the heat exchanger is increased so that the heat ratio is diminished.

- The rerouting of the solution counteracts this diminishing effect and yields an enlargement of the heat ratio by a certain absolute value (see the approximately initial straight shape of both arrays of curve, the solid and broken lines in Figure 5.14).
- The overlapping of temperatures by internal heat transfer from absorber to generator led to a reasonable enlargement of the heat ratio ζ with increasing degasification width.

The enlargement of the heat ratio is here outlined in Figure 5.14 for $p_o = 3$, $p = 7$ to 10, $T_o = -10°$ and $p_o = 1.5$, $p = 7$ to 10, $T_o = -25$ °C (solid- and broken-line arrays of curve). The degasification width is determined by the final concentration x_e of the solution in the generator—plotted on the abscissa in kg NH_3 per kg water—with an initial concentration of $x_i = 35$ %.

This effect continues to intensify when the temperature bands of absorber and generator approach each other with decrease of the pressure p with equal pressure p_o (as is shown in Figure 5.12).

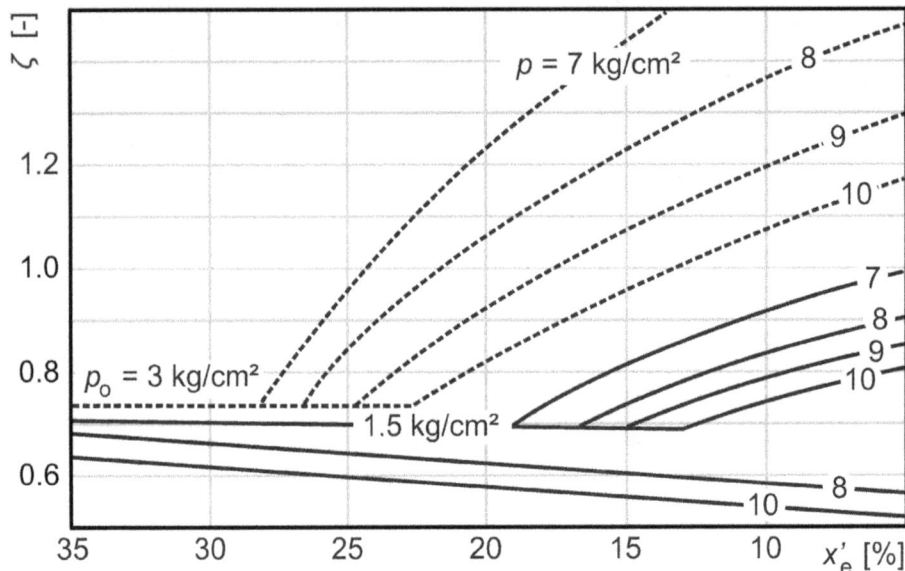

Fig 5.14: Enlargement of the heat ratio with overlapping of temperatures (with growing degasification width $\zeta_{outlet} - \zeta_{inlet}$) according to Altenkirch, 1929

The overlapping of temperatures in the cooling part of a resorption machine, which has here the effect of the generation of low temperatures (see Table 5.3), Altenkirch also wrote down in the claims 4-6 of his Patent[160] (including the solution prerouting in resorber and degasser).

The mentioned inequality of the specific heat consumption of the solutions is here *especially* large in the range of high working medium concentrations as they occur in the cooling part of the resorption machine.

So, for instance, at the pressure $p_o = 3$ bar of the degasser according to Figure 5.14 the weight ratio x of ammonia to water in dependence of the temperature is given by the following Table 5.3. Therewith the quantities of NH_3 per degree, here plotted in the line $\Delta x/\Delta T$, are degassed in these 3 intervals per kg water with the stated mean temperatures T_m. The same conditions are found in the resorber at the pressure $p = 7$ bar according to Table 5.3.

Thus, for instance, in the degasser between 10 and 22 °C on average released cooling of 0.04 kg NH₃ per K at 16 °C is not sufficient enough to discharge the heat of the resorption of 0.13 kg NH₃ per K released between 17.5 and 29 °C at the mean temperature 23.3 °C (a disproportion of 1:3).

Tab. 5.3: Iterative calculation of the heat consumption in the degasser and resorber

T [°C]	-2	10	22	32	17.5	29	44	56
x [-]	3.0	1.5	1.0	0.8	3.0	1.5	1.0	0.8
T_m [°C]	4		16	27	23.3		36.5	50
$\Delta x/\Delta T$ [-]	0.25		0.04	0.02	0.13		0.03	0.02

5.7.2 Adjustment due to Multiple Stages with a Narrow Solution Plot (Type I)

This demonstrated obviously that the means used according to 4.5.1 regarding rerouting and "overlapping of temperatures" with a too narrow solution plot (see the example according to Figure 5.15) still only yielded minor effects, especially when the boiling temperatures of the solution medium and the solution do not lie far enough from each other, for example the boiling temperature at the crystallization boundary of LiBr amounted only to 53 °C with the prevailing pressure over pure water at 0 °C so that the bridging range of temperature also amounted only to 53 °C, whereas with the ammonia-water mixture, on the other hand, the pure ammonia boiled at normal air pressure at -33 °C so that with 133 °C a more than twice as large range of temperature was available (see Figures 5.2 and 5.15).[45]

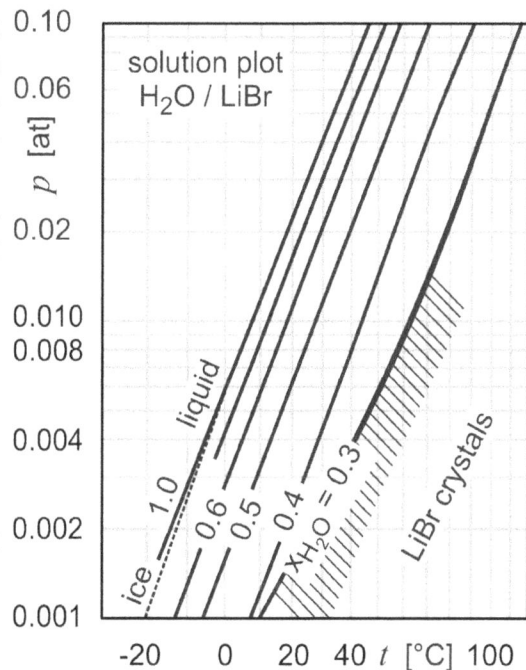

Fig. 5.15: Example of a narrow solution plot (H₂O-LiBr)

Similar to the cascade design outlined in the sections dealing with the thermocouple and reversible heating, even several absorption cycles could be combined with each other to form a cascade. This was managed in the same way as with *overlapping temperatures* by using the internal heat transfer, namely, by cooling the condenser and absorber of one stage with the generator of the following stage (*driving cascade, enlargement of the heat ratio*).

In order to reach low temperatures, absorber and condenser of one stage are to be cooled by the evaporator of the preceding stage. To solve such tasks, Altenkirch developed analogous equivalents of these dispositions, discussed each in the Sections 3 "Thermocouple for Reversible Heating", and 4 "Reversible Heating" at "Cascade Design". The following examples show the technical implementation according to Altenkirch:

---

[45] see WILHELM NIEBERGALL "Sorptionsmaschinen" in[103] figure 17.40

Halving of the Heat Output, Type I [46]

The method for utilizing a high slope of the driving temperature according to Figure 5.14 consists in the increasing of the pressure ratio by joining the several stages together with heat contact of the several stages whereby the degasification widths of all stages can be kept nearly equal. The external thermal conditions of the connection according to Figure 5.14 for *halving the heat output* shows Figure 5.21:

- Temperature of the desired cold production t_o = -10 °C.
- Temperatures of an available cooling medium: feed temperature $t_{CW,inlet}$ = + 10 °C, discharge temperature $t_{CW,outlet}$ = 30 °C.
- Temperatures of an available heating medium: feed temperature $t_{H,inlet}$ = +100 °C, discharge temperature $t_{H,outlet}$ = 80 °C

In the exemplary case for n = 2, the vapor which is produced in the generator between 100 and 80 °C is condensed in the condenser at 55 °C. The liquefying heat is supplied for heating through a heat exchanger cycle to the low-pressure generator of which vapor is again condensed in the low-pressure condenser with emission of heat to the cooling medium.

Therewith, the double amount of the refrigerant in the generator is available for the evaporation with only one heat supply. In the evaporator double the amount of cold is thus generated compared to a single-stage mode of operation, and the vapors are taken up from the absorber with heat emission to the cooling medium. In the train of the solution cycle the generators are connected in series due to a decompression upon the single pressure stages so that only one solution pump is required to hoist the rich solution from the absorber to the generator:

- Altenkirch's chosen connection has in the high pressure absorber, as the working diagram in Figure 5.14 at the bottom of the picture shows, in comparison to the generator about the double of the degasification width of the solution and thereby a correspondingly broad temperature band. This yields an improved adjustment to the Lorenz process, that is, in this case a further utilization of the cooling water by increasing its discharging temperature up to over 30 °C (see again Figure 5.12).
- Concomitantly, the connection leads to an avoidance of a solution pump (here it is again emphasized that Altenkirch was always attracted by the absorption machine especially due to its minor pump work—therefore its "nearly motionlessness").

Decrease of the Evaporation Temperature—Type I

In the exemplary case in Figure 5.16 the generator of the second stage "regenerates" the solutions enriched in both absorbers to reach low temperatures. Again, only one single solution pump is required [47], with which the rich solution is lifted from the low-pressure absorber into the generator.

---

[46] In order to distinguish the dispositions which perform different heat transformations, they are designated when working with internal heat transfer with "I", and otherwise, (see the following section) with "II".

[47] which deviated from the original schematic, and is added into Figure 5.16

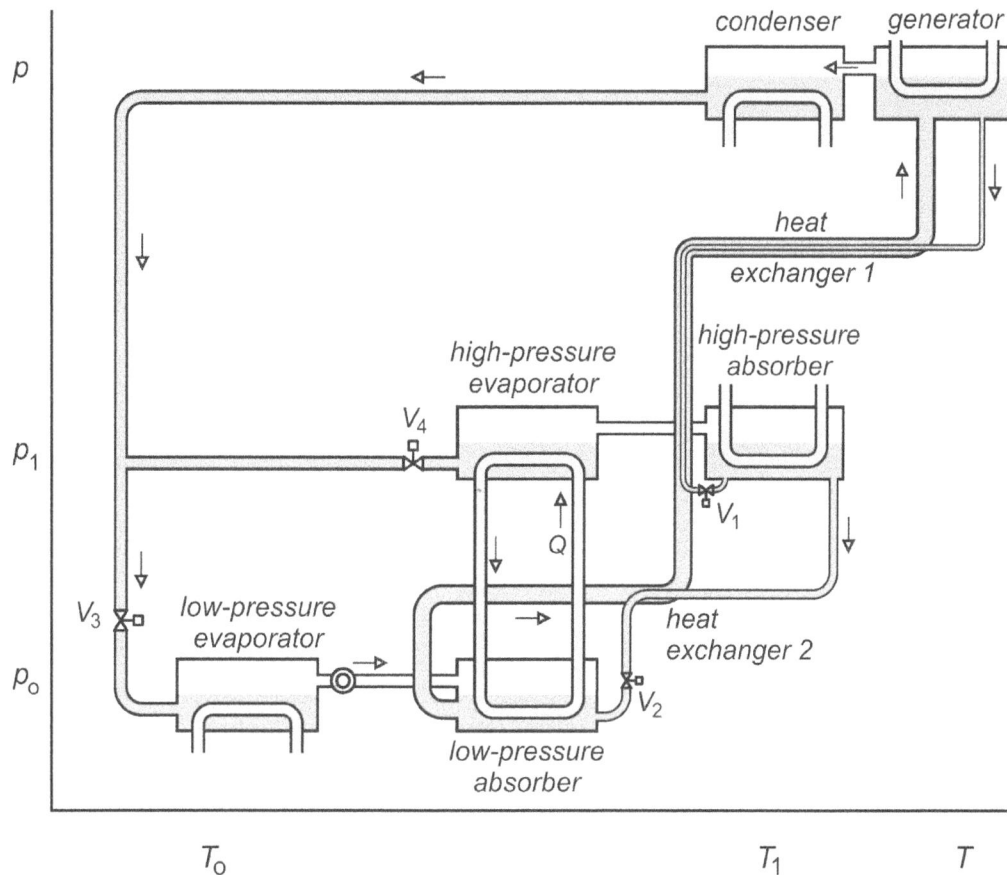

Fig 5.16: Schematic 'Low Temperatures I'

- The poor solution degassed due to the supply of thermal heat is precooled in the heat exchanger 1 in counter-current to the rich solution from the high-pressure absorber and is afterward decompressed via a control valve V_1 upon the pressure of the high-pressure absorber.

- After the uptake of the refrigerant from the high-pressure evaporator, the solution proceeds through the high-pressure heat exchanger 2 and from there after decompression in valve V_2 into the low-pressure absorber in which it is further enriched by uptake of refrigerant from the low-temperature evaporator.

- The refrigerant accrued in the condenser proceeds divided in halves after decompression either in control valve V_1 or V_2, and from there into the high- or low-pressure evaporator.

- The low-temperature cooling generated there, is produced with about the double expense of thermal heat from the generator—this underlies the thermodynamic relation in accordance with Equation 5.9.

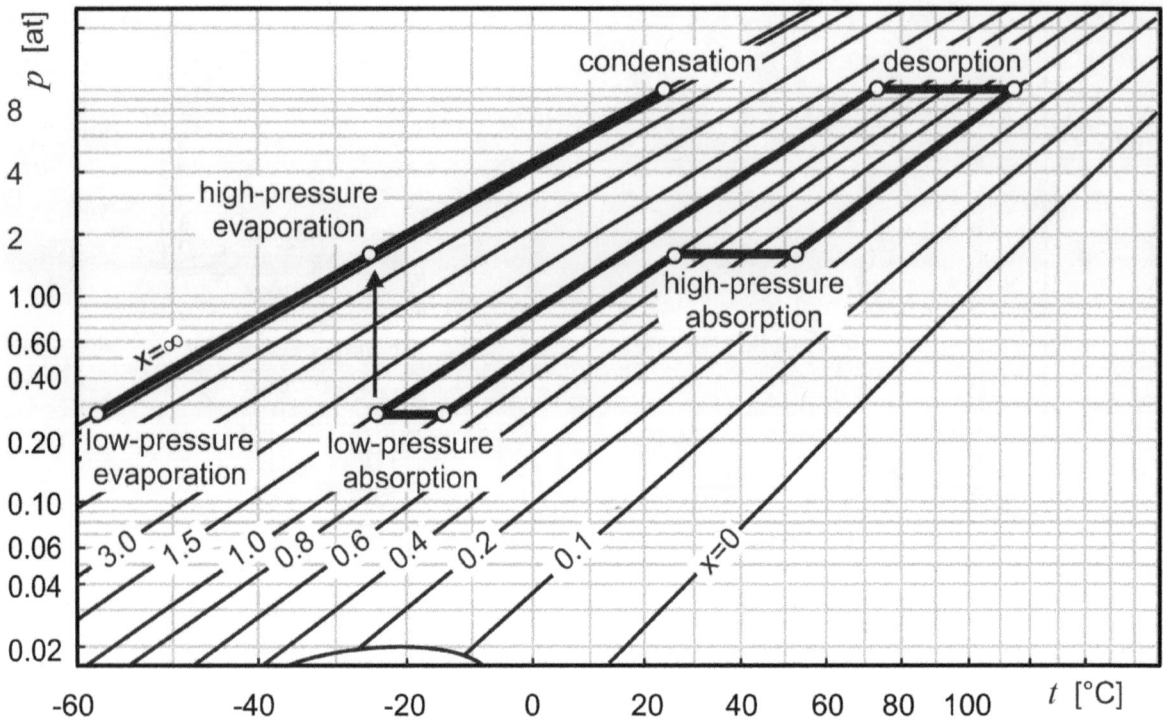

Fig. 5.17: Working diagram 'Low Temperatures I'

5.7.3 Adjustment by the Multistage Design to a Broad Solution Plot (Type II)

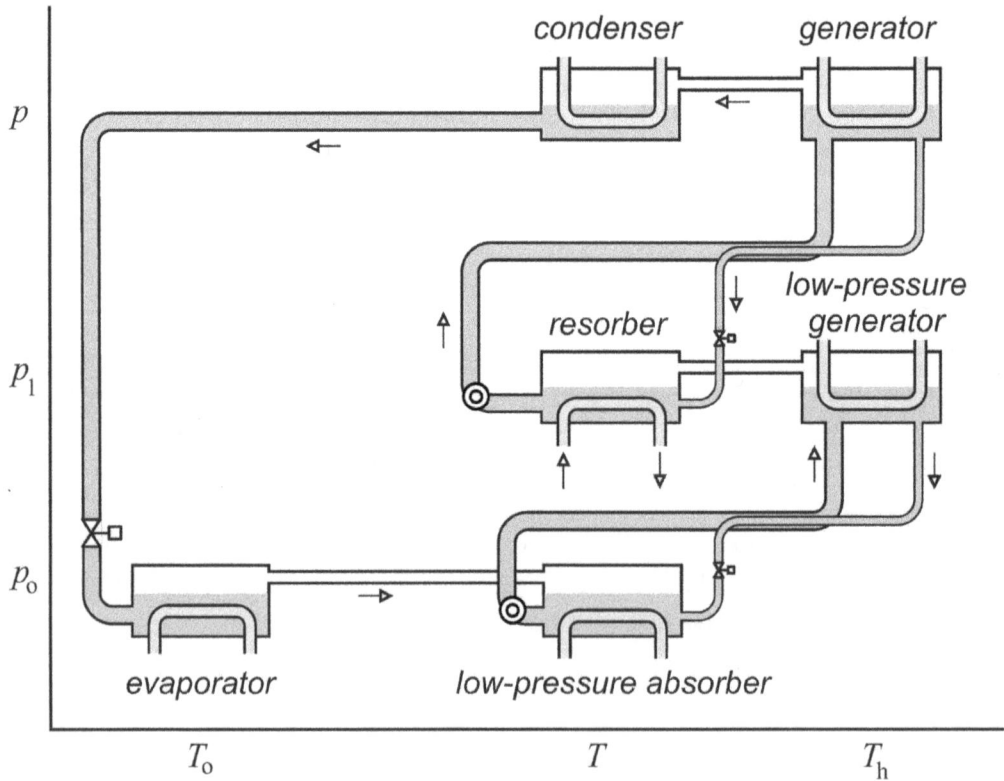

Fig. 5.18: Schematic 'Low Temperatures II'

Characteristic for the achievement of the thermodynamically necessary temperature differences (stroke/lift) of the driving or cooling part is the coupling of stages on the gas side with stronger extension into the solution plot (resorption).

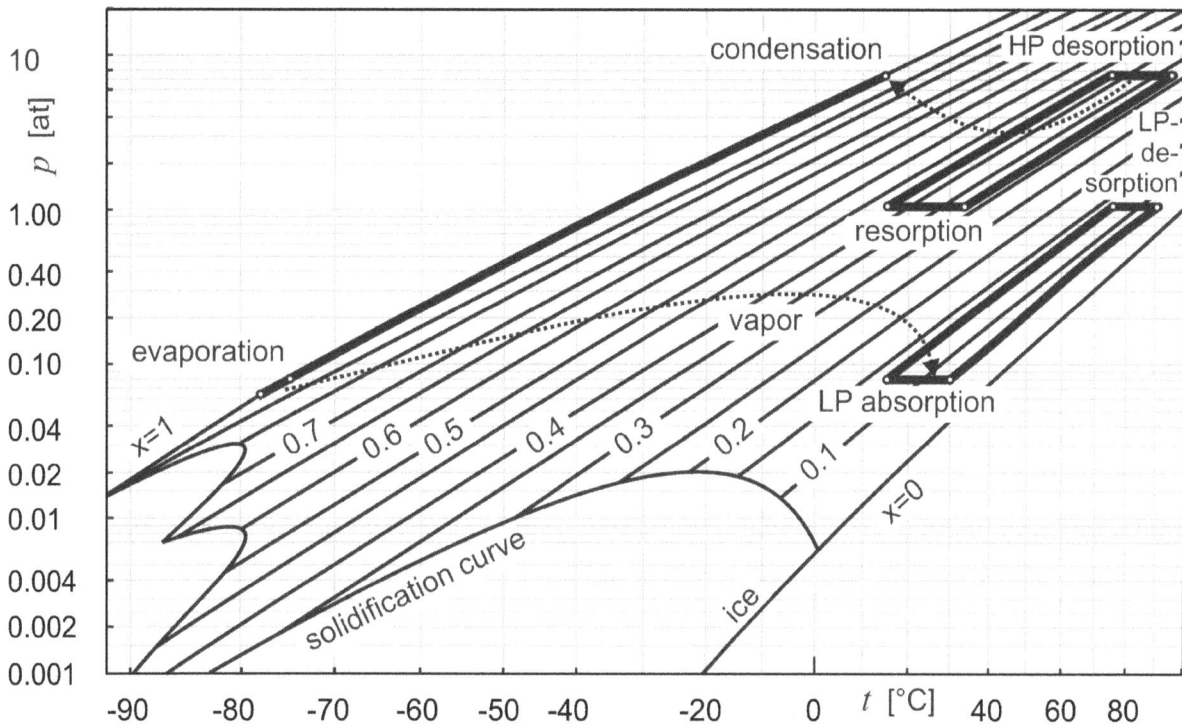

Fig. 5.19: Working diagram 'Low Temperatures II'

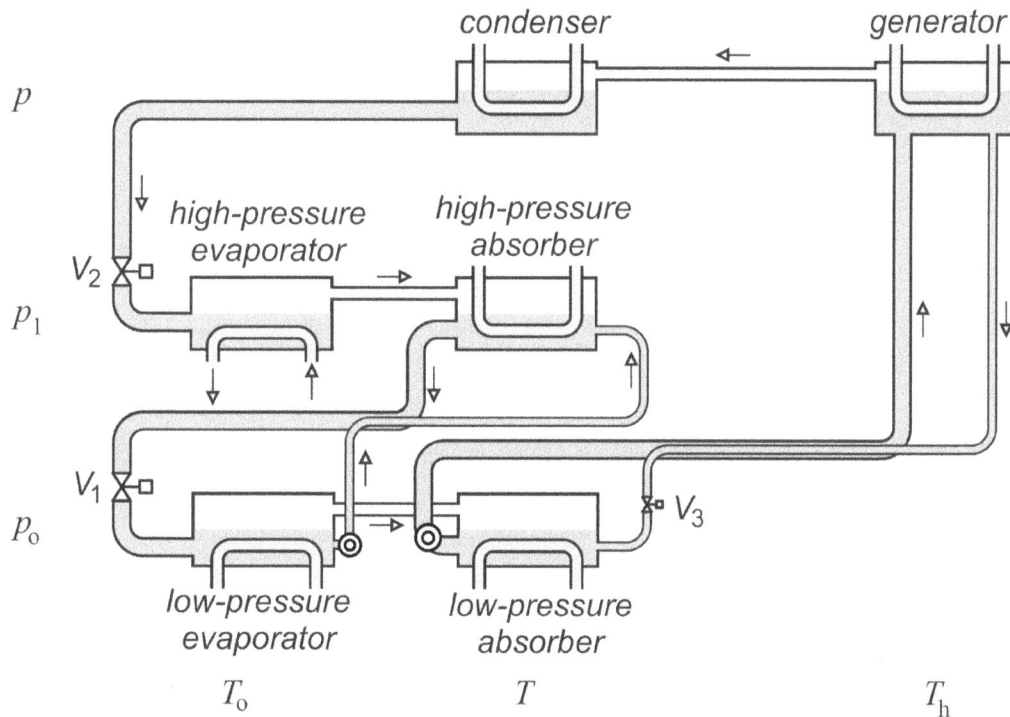

Fig. 5.20: Schematic: Increase of cooling by means of a two-stage cooling part

In order to gain an augmentation of stroke, the drive of the absorption machine similar to a multistage compression is caused by two (or more) compression stages connected in series (solution cycles) whereby the working medium vapor expelled in the low-pressure stage is absorbed in the absorber and expelled anew in the generator of the high-pressure stage—and then condensed in the condenser.

An example is the decrease of the evaporation temperature (type II) according to Figure 5.15 in the driving part due to a two-stage design.

The former cooling part with one-stage operational mode consisting of the condenser and evaporator can be designed as a resorption stage by reducing the mean concentration of the system. Due to the approximation to the LORENZ process it was at the same time possible to utilize the cooling media more effectively (the regeneration of the solution in the generator is herein denoted as "desorption").

Another example—outlined in Figure 5.21—is the increase of cooling by a two-stage design of the cooling part which also causes at the same time an enlargement of the heat ratio.

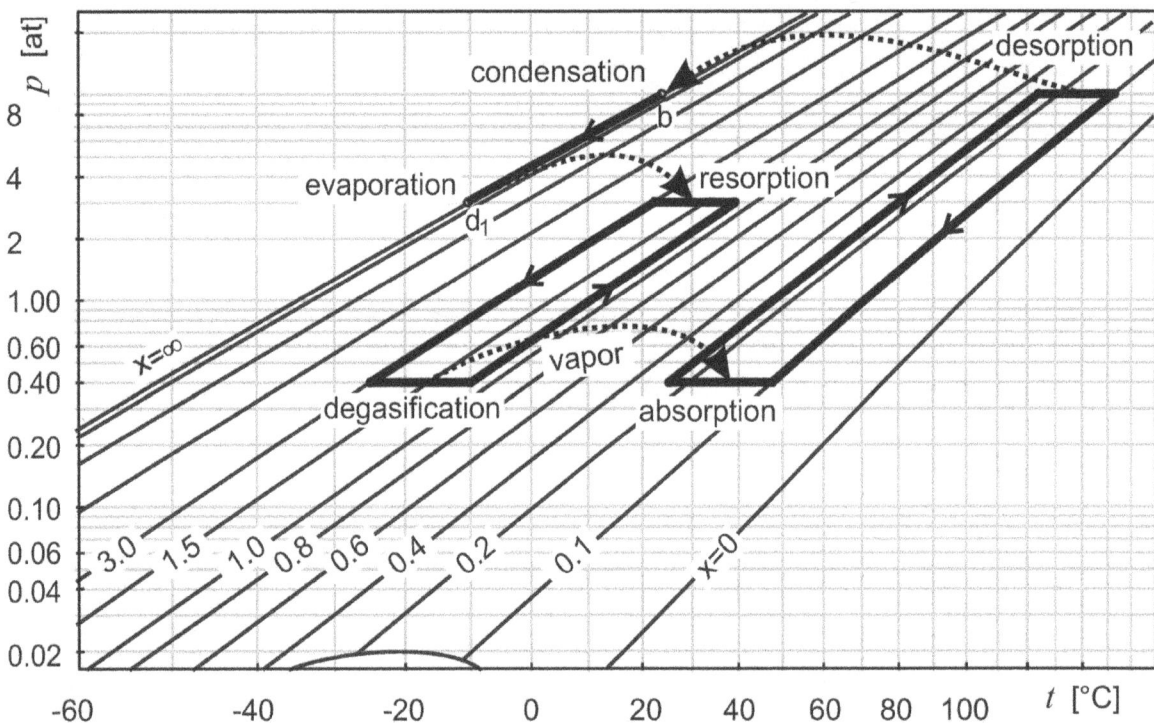

Fig. 5.21: Working diagram: Increase of cooling by means of a two-stage cooling part

As it is possible with a broad solution plot, the driving part can be executed with two stages in the outlined mode instead of having an internal heat transfer which always has a certain technical complexity.

5.8 All Improvements in an Overview

(1) The rerouting of the rich solution through the absorber and the poor solution through the generator leads to an improvement of the heat ratio depending on

the working conditions of the machine and especially on the degasification width with absolute values between 8 and 25 %[48].

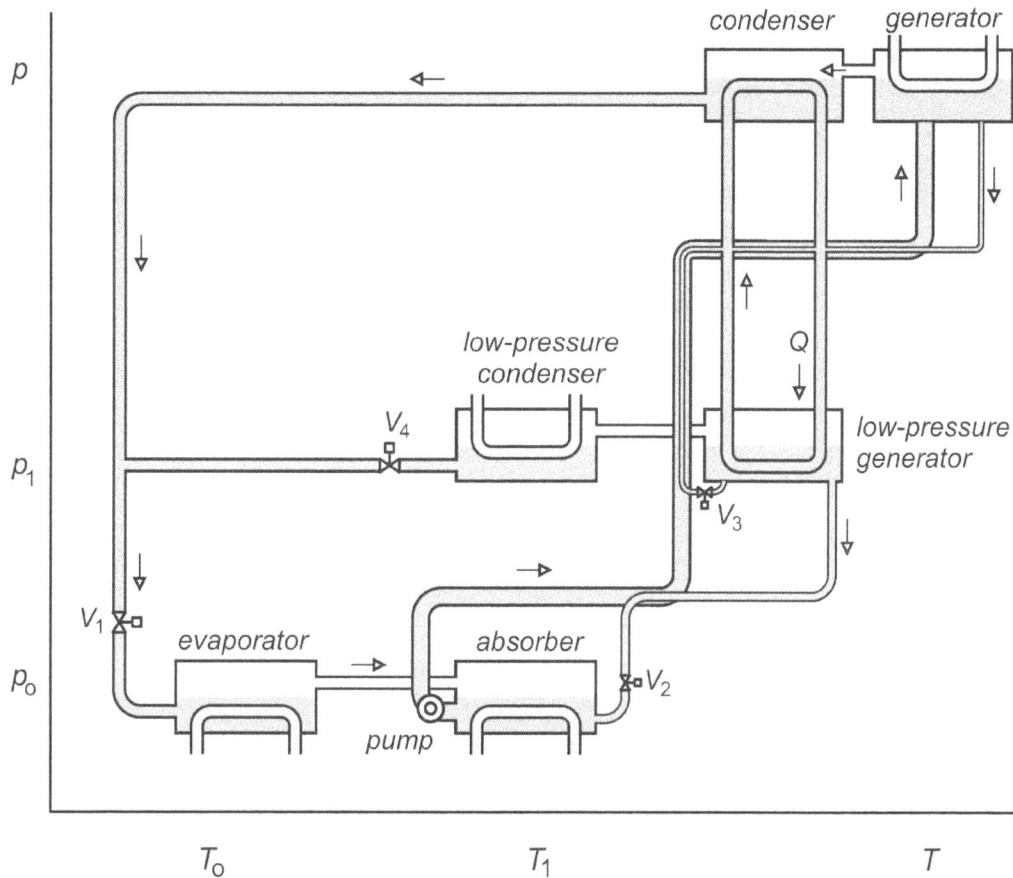

Fig. 5.22: Schematic for halving the heat output type I

(2) Due to the rerouting of the poor solution through the generator, the temperature difference between the poor solution and the preheated rich solution is reduced on the warm side of the heat exchanger. In this process the losses due to the inequality of the enthalpies of the rich and poor solution in the heat exchanger are decreasing. That means, among other matters (see also the following bullets), the rich solution approaches closer to its equilibrium state at the pressure p so that in this process the expelled vapor of the generator—with its operational mode in counter-current to the boiling solution—reaches an equilibrium state *with a higher working medium concentration (read: improved rectification!)* which is equivalent to the concentration of the rich solution.

(3) The thermal capacity of the vapor leaving the generator can be used in the vapor cooler for an additional preheating of the rich solution up to the boiling state of the generator at the pressure p[49].

---

[48] See separation column (rectifier) Figure 5.10, and FÖRSTER, H.[161] and STIRLIN, H.[112]

[49] See 5.10 [95] p. 8, and PLANK, R. [102] contribution by WILHELM NIEBERGALL

(4)   In addition, with the rerouting according to bullet 1 a part of the heat which otherwise is transferred to the rich solution for preheating in the heat exchanger is already supplied to it in the absorber (after increase in the pressure in the solution pump), which means relief provided to the solution heat exchanger and a diminishment of its losses.

(5)   The same is analogously valid for the generator. Only the extra costs for the *heating/cooling surfaces for these rerouting* are spent separately.

(6)   At the borderline case, when the temperature bands in the absorber and generator touch each other, as is managed due to the reduction of the circulating amount of solution, the *heat exchanger can be excluded entirely.* With further diminishment of the solution cycle an *overlapping of temperatures* takes place which can be used due to a constructive bringing about of a heat exchange between the hot part of the absorber and the cold part of the generator for saving on thermal heat for the generator—*heat ratios >1* can thus be achieved!

(7)   Due to the utilization of the *condenser/resorber heat for heating the generator* (like with the "cascading") *heat ratios >2* are to be achieved. For the implementation of the material pair ammonia/water "uncomfortable" high pressures must be taken into account. Altenkirch referred therefore to other, because of their flatter vapor pressure curves, more suitable absorption solutions, and stimulated their investigations[93].

(8)   The production for cold for the same heat output is doubled with the *application of the principle of resorption* in the cooling part with *multistage design— without heat transfer.* Heat ratios up to 1.2 had been developed. Here, there is a respective design of the company Borsig[95]. The method is applicable when the temperature of the cooling capacity has not to be very low. Where it remained over 0 °C, even two resorption stages, or one resorption and one condensation stage could be inserted between evaporator and absorber, see Figure 5.20 (heat ratio *nearly* = 2).

(9)   The *enlargement of the temperature lift* due to two-stage expulsion (here with two driving stages for the increasing of the pressure of the working medium vapor). This connection is also a multistage design without heat transfer. Herewith, facilities for dry ice generation[79] had been developed. Thereby, the temperatures in the generators can be kept that low that they are also able to manage it with "low value" waste heat (see Section 6 "Cryothermal Apparatus — A Motionless Heat Transformer", at "Some Advantages of the Cryothermal Apparatus").

(10) The degassing of the facility by washout of the exhausted air located at the point of the coolest and poorest solution which occurs in the process is particularly necessary for vacuum cryothermal apparatuses (see Section 6) to liberate the absorber from carrier gas under usage of a drip pump (see Section 9).

5.9   Response

Meanwhile the results of Altenkirch's work became common knowledge in the refrigeration and air-conditioning technology. A very interesting process involving an evaporator for optimal gas refrigeration in terms of the Lorenz process implemented by G. Maiuri[79], known by the term *diffusion absorption refrigeration machine* which, as with the machine by von Platen-Munters (Elektrolux[105]), attracted Geppert's[162] suggestion that led to pressure equalization between the high pressure of the generator/condenser and the low pressure of the evaporator/absorber by admixture of carrier gas. The diffusion refrigeration machine[62] enabled a broad and sliding temperature band to be generated in the evaporator—with a suitable measured intensity of the gas cycle—due to an achieved extended slope of the partial pressure of the working media in the circulating carrier gas (hydrogen/helium).

This wide range of options for coupling and combination of absorption and resorption stages in driving and cooling parts as well as in connection with compression and working machines are deployed for the optimization of a variety of operational and energy-relevant tasks[86].

Universal *systematizations of multistage designs* have been carried out by Nesselmann[84], Alefeld[47] and, above all, by Ziegler and Alefeld[129]. In the latter paper easy computing methods are submitted for projection and quick comparison of multistage designs of absorption cycles as well as the *modeling*[48] of the operational behavior of one-stage cycles, specifically:

- the sorption processes and their heat transfers to external carrier media for cooling, heating and refrigeration,
- their heat coupling with the sorption processes of other cycles according to temperatures, mass flows, and heat transfer capacities, and
- the heat transfer processes in the heat exchangers themselves.

This succeeded by introducing compact criteria which enabled a rapid computer simulation in good approximation. For that the following one-stage cycles with condensation (or resorption) are distinguished:

- heat pump (or refrigerating machine, respectively) with counter-clockwise orientation of the solution cycle,
- heat transformer with clockwise orientation of the solution cycle.

The assigned heats are Q_D $(=Q_o (= cold))$, Q_G (= thermal heat of the generator), Q_C (waste heat of the condenser), Q_A (= waste heat of the absorber), $Q_C + Q_A = Q$ (sum of the waste heats of condenser and absorber.) The simplifying assumptions for the relations of the heat flows at a stage with regard to the revolving working medium quantity of one cycle are

$$Q_C = Q_o \quad \text{and} \quad Q_G = Q_A \qquad \text{(Eq. 5.30)}$$

In the case of refrigeration the heat ratio is interesting

$$\zeta_o = \frac{Q_o}{Q_G} \qquad \text{(Eq. 5.31)}$$

With coupled cycles (see the example according to Figure 5.24, upper part of the picture) the waste heat $Q_{C,a} + Q_{A,a}$ of a cycle a heats the generator of a cycle b

$$Q_{G,a} = Q_{C,b} + Q_{A,b} \qquad \text{(Eq. 5.32)}$$

With the simplifications according to Equation 5.30 and the variables m_a and m_b of the working media streams of the cycles a and b this becomes

$$m_{\mathrm{b}} \cdot Q_{\mathrm{G}} = m_{\mathrm{a}} \cdot Q_{\mathrm{C}} + m_{\mathrm{a}} \cdot Q_{\mathrm{G}} \quad \Leftrightarrow \quad \frac{m_{\mathrm{b}}}{m_{\mathrm{a}}} = \frac{Q_{\mathrm{C}} + Q_{\mathrm{G}}}{Q_{\mathrm{G}}} \qquad \text{(Eqs. 5.33, 5.34)}$$

Thus the heat ratio of a two-stage design increases therefore up to

$$\zeta_{\mathrm{a+b}} = \frac{m_{\mathrm{b}} \cdot Q_{\mathrm{C}}}{m_{\mathrm{a}} \cdot Q_{\mathrm{G}}} = \frac{Q_{\mathrm{C}} + Q_{\mathrm{G}}}{Q_{\mathrm{G}}} \cdot \frac{Q_{\mathrm{C}}}{Q_{\mathrm{G}}} = \left(1 + \zeta_{\mathrm{o}}\right) \cdot \zeta_{\mathrm{o}} \qquad \text{(Eq. 5.35)}$$

In the upper part of the Figure 5.24 the heat flows of the coupling, and the heat flows *from* and *to* exterior heat carriers are outlined with directed serpentine lines. In the lower part the depiction of the interior transfer processes is omitted.

For an *evaluation of irreversibilities* inside a defined absorption machine process, see S. Unger[116], [117].

The application of heat transformations with the *concentration* of solutions due to vapor compression (heat pump)—with a nearly reversible process conduct—which enables also reversely the storage of the working capability for a compressor (e. g. due to operation with a cost-effective night current), is described by H. Voigt[119]. Thereby, this (concentrated/poor) solution can be used again—in time-shifted operational mode—for the aspiration of the working medium for the operation of a refrigerating machine or heat pump. This application was also investigated by Voigt, H. who demonstrated the economic benefits. In 1983 G. Alefeld reported similar options for design connections [47]50.

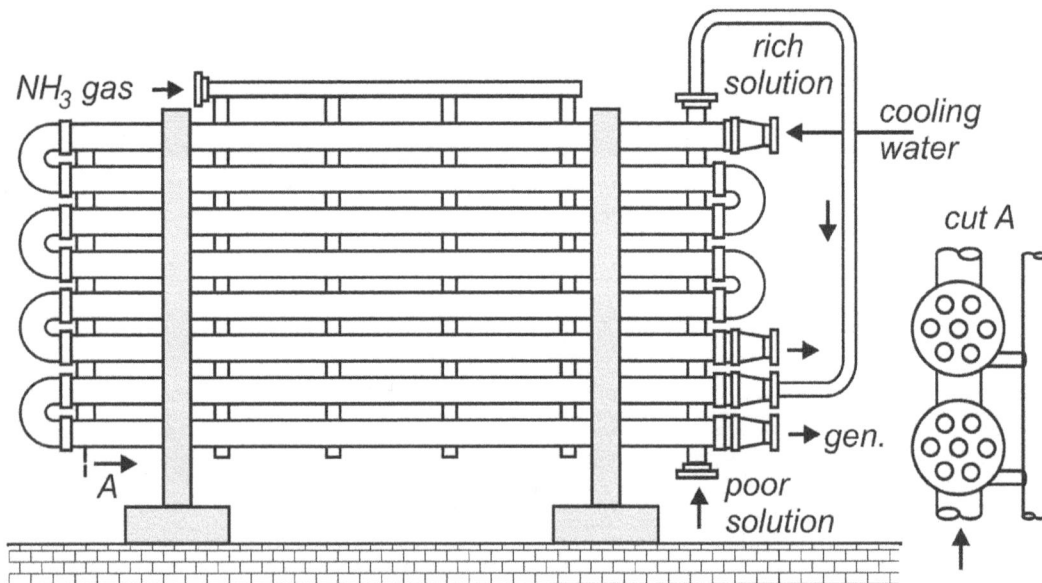

Fig. 5.23: Multi-bundle pipe absorber of the company Borsig *Berlin*

Some of Altenkirch's innovations regarding the absorption machine the Berlin company Rheinmetall-Borsig introduced into practice but only at the end of the 1920s when Altenkirch had already worked for several years (from 1920 – 1925) as a scientific collaborator for this company!

---

50 see p. 225 et sequ. in [44] and Chapter 2 "Absorption Aggregates"

The first two-stage absorption machine was set up 1929 at the Technische Hochschule Karlsruhe and was thoroughly investigated. Two-stage absorption refrigeration machines for 40,000 kcal/h ($\approx$ 46.5 kW) for two different cooling sectors from -10 and -30 °C at the Städtisches Gaswerk[51] Berlin-Charlottenburg go back to the year 1930—a remarkably large heat ratio was achieved[90]. In the following years the company Borsig built a large number of one- and two-stage facilities according to Altenkirch's connections. From 1930 until 1945 the cooling performances became steadily larger.

An absorber with rerouting of the solution is shown in Figure 5.23. This multi-bundle pipe absorber has a water-cooled part and a solution-cooled part (the actual rerouting of the solution), as well as a gas distribution system. The poor solution enters on the bottom side and absorbs on its way from the bottom to the top the laterally entering ammonium gas, where it enriches itself. The rerouting of the solution has been directed into the initial part of the absorber and takes there the heat up from the absorber of higher temperature. In this course the rich solution is preheated for the entry into the heat exchanger (the solution pump at the upper exit of the absorber has been omitted in this picture)[52]. Moreover, the Maiuri Refrigeration Ltd. operated until 1939, while being in touch with Altenkirch from London, in England, France and the United States their famous facilities for production of solid carbon dioxide under co-use of the designs developed by Altenkirch. Worthy of remark is here the intended rectification of the generator vapors which is very important for deep-freezing facilities and is strongly favored by Altenkirch's rerouting process through the generator because the rich solution from the absorber is conveyed at a cool temperature into the generator and enables there a strong intensification of the rectification of the generator vapor with counter-current conduct to the solution.

In modern times the multistage (double effect, triple effect ..., etc.) designs for the enlargement of the heat ratio received attention in the United States. Due to realistic computations F. Ziegler, R. Kahn, F. Summerer and G. Alefeld[127] investigated the utilization of the given input temperature differences for the enlargement of the heat ratio by using the multistage design (i. e. with "multi-effect absorption chillers"). There, with not too high requirement for the temperature lift, an improvement of the heat ratio up to a value of 2.6 until the 5th effect (stroke) (5-stage design) is still achieved.

In the recent relevant literature Felix Ziegler's summarizing paper "Sorptionswärmepumpen"[53] sets out the significant fundamentals of Absorption Refrigeration Technology[125] in a very compact form.

Ziegler[131] reports on the development of an absorption refrigeration machine for smaller performances (around 10 kW) for which there are applications in the field of cooling powered both by solar energy and low temperature waste heat of, for example, cogeneration plants.

---

[51] Municipal gas-plant [translator's note].

[52] A connection from the exit of the rich solution until the mid-level of the horizontal assembled cooling tubes of the absorber—which are analogously further flown through by cooling water—are supplemented. The solution pump in the stream of the rerouted solution is here omitted due to space. The rerouted solution leaves the underneath horizontal tube into the right direction and travels to the generator.

[53] "Sorption Heat Pumps" [translator's note].

For the successful use of the absorption refrigeration technology, reliable computation methods should be available, also especially for multistage designs. After submission of a systematization of different modes of designs by Nesselmann and particularly by Alefeld, substantial progress had been made by modeling both the state diagram (pressure/concentration diagram, Dühring-Plot) and the thermal effects through i-ξ diagrams due to material values like solution heats, heat capacities in heat exchangers, the mode of processing (cooling stage, heat pump stage) for a meaningful computer simulation of the internal and external effects (heat transfer between sorbing solution and exterior media in heat exchangers including their temperature changes)[127].

Niebergall reports on special material pairs for special applications of heat transformation. There are new research findings for computer-assisted calculations, experimental investigations, and computerized simulations of processes of absorption machines. In one example there is described a facility for raising of waste heat, for example for drying and distillation processes (fruit juice production, xylene distillation)[111].

Ziegler[126] refers to new media which can be added to the solution to increase the sorption speed in these devices.

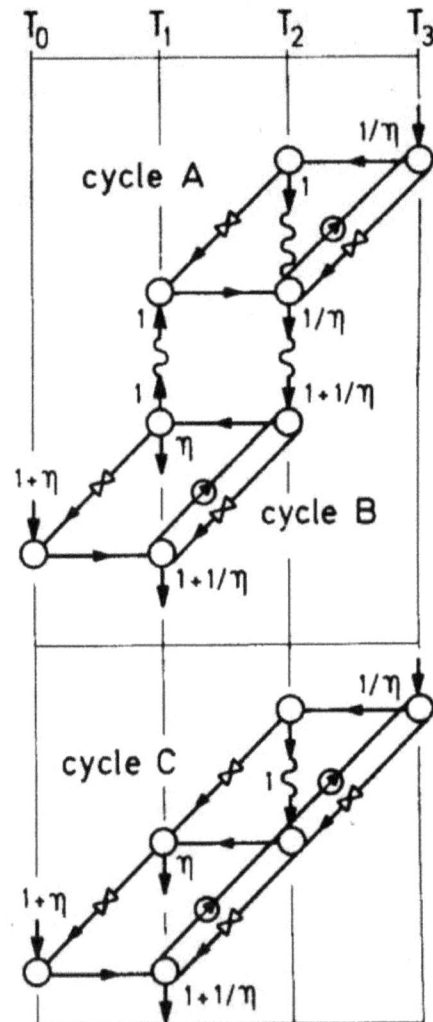

Fig. 5.24: Two-stage cascade according to Ziegler, F. (1997)

6 The Cryothermal Apparatus—A Motionless Heat Transformer

This section covers Altenkirch's further development of the absorption machine to a wear-resistant machine type with cost-effective construction without the need of mechanical moving parts (as of 1920).

6.1 The Principle

The characteristic part in Altenkirch's patent specification[137] with the title, 'Absorption machine with two vessels of different pressure in which one takes over the absorption and the other the degasification', is as follows, '... characterized by a liquid column in this tube which conducts the liquid from the vessel of low pressure (absorber) into the vessel, spatially arranged in lower position, of higher pressure (generator) and which maintains the pressure difference in these both vessels during operation.'

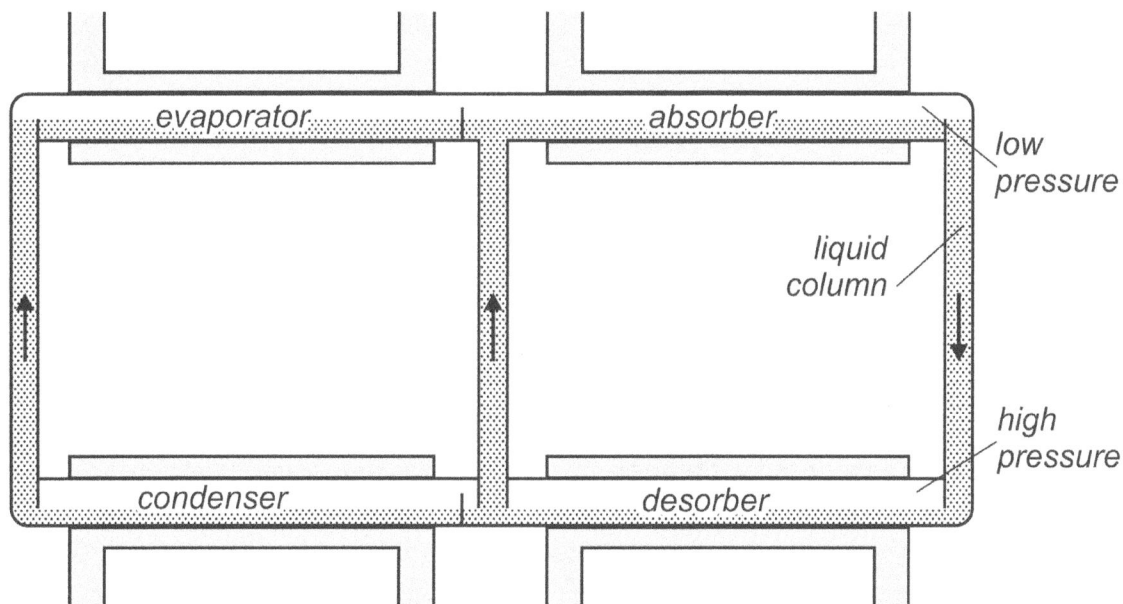

Fig. 6.1: Absorption machine according to Altenkirch[137]

The patent for the cryothermal apparatus is historically the first one of its kind. It claims the patent for the *liquid column in this tube which conducts the liquid from the vessel of low pressure into the vessel of high pressure*.

The required lifting system for real models of the machine to overcome the slope height in sorption apparatuses with trickle stretches is described in the successive patent[140] with the title '*Absorption Machine*.' It claims the right to the idea of the gas bubble pump (*boiling tube* in Figure 6.2) in which the gas bubbles of the expelled vapor on their way to the gas separator carry the poor solution boiled-out in the generator up into the gas separator (thermosyphon effect).

In claim 4 the patent focuses on the elimination of the disturbing vapor bubble development from the degassing hot, poor solution in the generator due to its hydrostatic lowering of pressure when *ascending to the absorber.* This will be enabled due to the cooling of the solution in the heat exchanger 2[54].

In order to keep the overall height of the machine in the centimeter range, the height of the liquid columns may not extend above a certain point. This was easily done with water as refrigerant although Altenkirch initially used the sulphuric acid/water solution as an absorbent that was later substituted with the less aggressive lithium chloride or lithium bromide/water solutions.

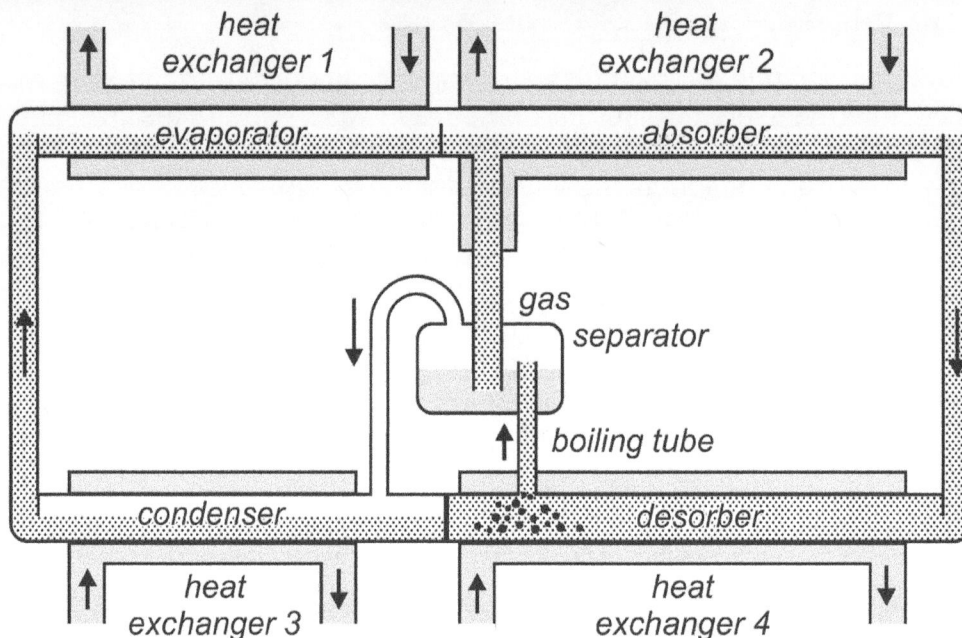

Fig. 6.2: Principle of the cryothermal apparatus with a lifting system for circulating the solution

Measures to guarantee a stable thermo-hydromechanical operation when boiling in vacuum due to the vapor bubble transport for lifting up the poor solution—by Altenkirch in a preferably as a simmering coil constructed conveying tube—are also mentioned in claim 8 of this pioneer patent (buffer space in the supplying pipe for the simmering coil). This claim describes moreover the rerouting of the solvent-containing residual solution from the evaporator into the absorber through capillary 1 in Figure 6.3.

Altenkirch coined the German term 'Kryotherme' (cryothermal apparatus)[55] to describe an absorption machine with full or partial pressure equalization due to liquid columns whereby the solution was rotated by means of a vapor bubble pump according to the just mentioned thermosyphon principle. A technical description of the functioning is shown in Figure 6.3:

---

[54] This problem virtually does not exist in absorption machines with pumps and valves since the poor solution generally can be sufficiently pre-cooled in the heat exchanger before it expands in the throttle valve.

[55] The English term "cryothermal apparatus" is used for Altenkirch's German term "Kryotherme" according to Karl Stephan's usage in the International Journal of Refrigeration (1983) [translator's note].

. To avoid the liquid in the combined generator/solution transport system not only being driven into the gas separator but also thrown back into the absorber due to the boiling impact, *a liquid buffer* is inserted into the absorber drain pipe for the rich solution in front of its opening into the generator simmering coil which itself is connected via an upward conducting pipe with the g*as buffer.*

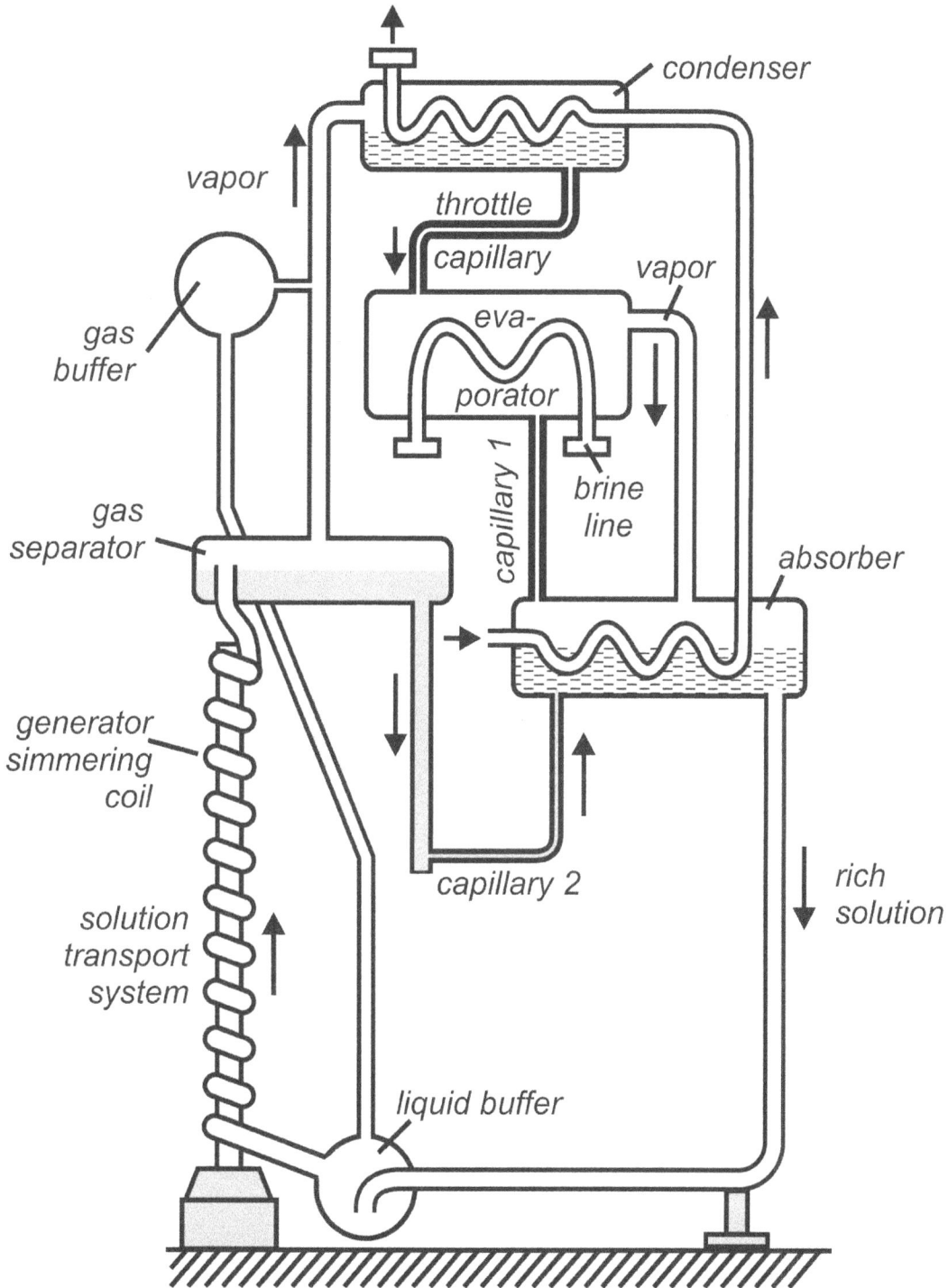

Fig. 6.3: Cryothermal apparatus with a solution cycle by means of vapor bubbles (thermosyphon effect) according to Altenkirch, E.[140]

- The absorber drain pipe (*rich solution*) is bent downward inside the liquid buffer so that quantities of gas which can retrogress from the solution transport system do not penetrate into the absorber drain pipe. They rather climb the upward leading pipe to the gas buffer which is connected with the vapor line (*vapor*) that runs from the gas separator to the condenser. In this way the quantities of gas may convey into the condenser.

- For prevention of the immediate transfer of vapor from the generator to the absorber, the connection from the gas separator to the absorber is formed by a liquid column (capillary 2).

- From the condenser which offers a certain space for a supply of refrigerant, the condensate enters the evaporator through a *throttle capillary*. Through the *brine line* flows the medium to be cooled.

- The liquefied refrigerant leaves the condenser—initially with an infinitesimal pressure difference—through the *throttle capillary* into the evaporator due to the liquid column being effective by the height distance. The throttle capillary prevents at the same time the snap-through buckling of the vapor from the condenser into the evaporator since the capillary has much more resistance to the vapor than to the liquid.

- It must be expected that small quantities of the solvent are transported along with the working medium into the evaporator. At the end of the evaporation process the concentration of the solution in the condensate is the highest. At this point, the residual liquid is rerouted through the *capillary 1* back into the absorber for cleaning the evaporator.

With the cryothermal apparatus a new quality of absorption machine was created without maintenance-requiring mechanically moving parts like pumps and valves.

Due to the required overall height when operating with pressure maintenance by liquid columns for absorption machines in which the working medium is liquefied and re-vaporized (so called *condensation-type absorption machines*) virtually only the known material pairs water as working medium and sulfuric acid or lithium bromide as solvent came in to question at that time[56].

Altenkirch's idea of a (reversible) *cooling aggregate*, solely operating with heat and therefore with low maintenance, filled exactly the technological niche for the development of the refrigerator, but he had a long-term vision for creating cheap units of heat transformation with manifold technical applications.

6.2   The Problem of Overall Height / Atmospheric Resorption

In order to be able to maintain the successful material pair ammonia/water Altenkirch used later (after World War II when the above mentioned Elektrolux patents had expired) the principle of resorption by introducing a resorption cycle through condenser and evaporator also due to a combination with a carrier gas admixture[57].

---

[56] but note the subsequent Section 6.4.2

[57] For this purpose Altenkirch referred to GEPPERT's proposal (1900). According to this the pressure difference between condenser/resorber and generator, on one hand, and evaporator/degasser and absorber, on the other, could be equalized by admixture of carrier gas. This type of system became, however, first practically functional by permitting a flowing movement through absorber and evaporator. This could be reached by different measures (see further below).

Regarding the usage of the material pair ammonia/water, the requirement for lowering the pressure due to resorption in the generator/absorber cycle led to a higher water content of the solution so that the rectification of the generator vapor gained importance. In return, as mentioned above, the rerouting of the solution in the sorption apparatuses especially in the generator created the optimal opportunity for using the counter-current flow of the expelled vapor in heat contact and mass transfer to the solution stream inside the generator for a reversible rectification (see Section 5).

The solution cycle in the resorber/degasser system Altenkirch brought about through a lateral passing of the generator vapor into the line of the poor solution which came from the degasser whereby this solution was lifted up into the resorber according to the *principle of air-lift pump (mammoth pump)*[53].

Regarding the limitation on cooling temperatures above 0 °C the use of water as refrigerant naturally had advantages since the overall-height problems with the intended maintenance of pressure differences inside the machines due to liquid columns could completely be minimized by the slight vapor pressure of the water. Therewith the complication of an additional resorption cycle for reducing the pressure inside the machines as well as the complication of a partly pressure equalization by carrier gas was omitted.

6.3   Some Advantages of the Cryothermal Apparatus

The cryothermal apparatus (according to Altenkirch's system)[95], as it was described in two patents[137],[140], achieved a heat ratio more than a double, compared with the admixture of carrier gas.

NIEBERGALL mentioned, inter alia, that in the United States these machines could be placed on upper floors of buildings due to their low weight and vibration-free and quiet operation.

Regarding the conveyance of the solution according to the thermosyphon principle compared with hydromechanical pumps it was expected that the transport would be much more robust against disturbances which occurred because of occasionally entrained dirt or material particles. *Thus,* Altenkirch *had in mind to manufacture a cryothermal apparatus of which the operational process was near the atmospheric pressure in order to use inexpensive building materials such as stonework which could be integrated into buildings—because the mechanical stress of the vessel walls due to the internal pressure would be negligible* (see below).

In order to counteract the entrainment of cold to warmer parts of the system and, vice versa, the protraction of heat into the evaporator, a gas heat-exchanger was inserted through evaporator and absorber into the stream of the circulating carrier gas.

According to the invention by Platen-Munters (Elektrolux), the hydrogen was usefully used as carrier gas whereby the propulsion of the gas cycle came about due to differences of the average molecular weight between the poorer solution gas column of the absorber and the richer solution gas column of the evaporator.

Due to an exposé by Altenkirch (see Section 9.4.1), the heat ratio was, however, considerably second place with one-half to one-third behind that of his system with maintenance of pressure by liquid columns, which he developed for the company Siemens-Schuckert-Werke with or without the use of the principle of resorption in the years after 1920.

Of course, material pairs had to be established which aside from their thermal appropriateness also guaranteed the required safety. The operating process must therefore take place once more at atmospheric pressure or at low pressures. The equalization of pressure up to the atmospheric level should take place according to Altenkirch with air as carrier gas by gas cycle propulsion due to molecular weight differences between ammonia and air. The resorption cryothermal apparatus required however two liquid cycles, but had the advantage of the variability of the pressure range. With the decrease of the working medium concentration and water as solvent a careful rectification of the working medium vapor was required. The resorption machine in combination with pressure maintenance due to liquid columns and partly pressure equalization due to admixture of carrier gas was considered an optimal solution.

6.4   Response to the Developments of the Cryothermal Apparatus

6.4.1   Vacuum Cryothermal Apparatus

Altenkirch reported that the first field of application was the refrigerator in which the company SIEMENS-SCHUCKERT-WERKE had a stronger interest than the company BORSIG. Therefore, Altenkirch signed a contract with the company SIEMENS-SCHUCKERT-WERKE in accord with BORSIG. The development led to the construction of cryothermal apparatuses with the material substances water and sulfuric acid. Soon some trial machines which were built glass operated with best success.

Later, in the United States, machines were developed by the companies SERVEL INC. and CARRIER CORP. with iron as building material using the material substances water/lithium chloride or water/lithium bromide and which were used for instance for the operation of air-conditioning systems.

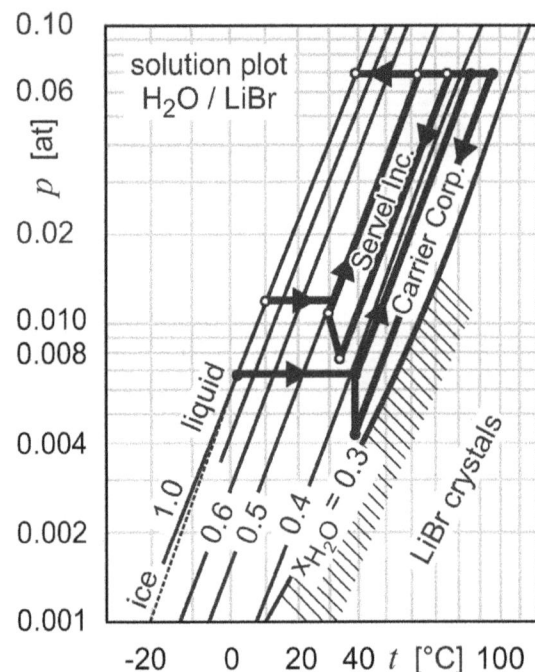

Fig. 6.4: Operating processes of vacuum cryothermal apparatuses with the material substances $H_2O/LiBr$

There are results about that development in the specialized literature[58]. The American company Electrolux (SERVEL CORPORATION)[59] also used for larger air-conditioning systems up to 15,000 kcal/h (ca. 17.5 kW) liquid columns for the maintenance of pressure differences.

---

[58] for example, in "Z. d. Ver. Dt. Ing. 91 (1949)," pp. 493-496 and [95]

[59] also cf. http://www.roburcorp.com/company/history/history.html; retrieved 2013-08-09 [translator's note].

6.4.2 Systems with Pressure-Equalizing Carrier Gas / Diffusion Machine

The equalization of the pressure difference between evaporator/absorber, on one hand, and generator/condenser, on the other, was also possible—as mentioned—by means of carrier gas according to GEPPERT, H.[162] whereby the cold production is achieved by the principle of evaporative cooling in the evaporator (e. g. as takes place in cooling towers with re-cooling of the cooling water of steam power plants), meaning that the regular evaporation of the refrigerant in the evaporator is replaced by an evaporation[60] into the carrier gas. Due to the uptake of the cooling medium vapor by the carrier gas the cooling medium vapor's partial pressure in the carrier gas increases and must be "washed out" of the carrier gas by the absorption solution in the absorber again. For this purpose the diffusion velocity of ammonia into the carrier gas is co-determining.

6.4.3 The System According to von Platen and Munters

The Swedish researchers VON PLATEN and MUNTERS[105] introduced into this cold production system a movement of the carrier gas in a cycle between evaporator and absorber whereby they used propulsion due to molecular weight differences (see further below the caption to Figure 6.5).

The course of temperature in the sorption apparatuses received an additional degree of freedom by introducing the carrier gas (see the articles on the diffusion absorption machine[61]). The partial pressures of refrigerant and carrier gas add up to the pressure total in the machine[62] whereby the pressure of the pure refrigerant in the generator/condenser is fixed, which is determined according to the nature of the refrigerant itself (example ammonia) in the condenser and the external cooling temperature (water or air).

The use of carrier gas instead of liquid columns allows the favorable application of effective refrigerants such as ammonia and thus limits the overall height of the machine in essence only to the vertical extension of the sorption vessels required for the absorption metal trickle sheets.

A disadvantage of the carrier gas application is the necessity to insert a gas heat exchanger to pre-cool the condensate from the condenser together with the warm and poor gaseous mixture coming from the absorber in counter-current to the cold and rich mixture coming from the evaporator (cf. Section 5.4.2, Cycle of the Working Medium in the Absorption Machine, second Paragraph). As already mentioned earlier, losses occur due to imperfections (particularly due to the longitudinal heat conductance) of the heat transfer—also the constructive complications due to the striving for high effectiveness of the heat transfer are considerable.

Thus the heat ratio compared to Altenkirch's system with liquid columns is accordingly reduced (see footnote [63]).

---

[60] Since the evaporation underlies the diffusion of gases into gases the refrigeration machines which are based on this are also named diffusion machines.

[61] Maiuri, G.[79] developed applications for generating low temperatures (dry ice) by powering the carrier gas cycle with a fan and using a common hydromechanical solution pump.

[62] except small differences in pressure due to slope heights in the sorption apparatuses (see the caption to Figure 6.5.)

The Swedish researchers used a thermosyphon pump as a solution pump by Al-tenkirch[63] to surmount the slope heights (trickle heights) in the absorber but he had for this the priority—the result was a tough competition of patents.[64]

Figure 6.5 shows the principle of the Electrolux cryothermal apparatus, which consists of the main components: absorber, thermosyphon pump, vapor cooler and condenser, and evaporator with liquid and gas heat exchanger.

Due to the choice of hydrogen (molecular weight = 2) as carrier gas and ammonia (molecular weight = 17) as refrigerant a high driving difference for the gas cycle and again, simultaneously, due to the low molecular weight of hydrogen, a favorable, high diffusion rate for the diffusion of ammonia within hydrogen was realizable[65].

Caption to Figure 6.5:

. The refrigerant vapor generated by heating the solution in the thermosyphon pump 8 (coil and upward outgoing tube) pushes (during the accelerating phase of the machine or with the changing pressure total, respectively) the cooling medium/carrier gas admixture from the condenser in different extents back into the condenser's end so that the admixture can be liquefied.

. The solution which is initially *rich* in refrigerant is dragged along inside the thermosyphon pump due to the ascending gas bubbles of the generated refrigerant vapor into the annular space 7 of the generator which consists of a gas separator. The generator vapor passes through the somewhat wider rectifier tube, which is inside provided with metal trickle sheets, in counter-current to the solution to be washed out there (which is very effective with a solution rich in water content). This *poor* refrigerant solution leaves then through tube 9 the jacked space of the generator and the jacked space of the *solution heat exchanger.* It travels to the

---

[63] see. Figure 6.3 and [140]

[64] See Altenkirch[45]: From there the summarized citation: In the same year, the "liquid cycle by gas bubble transport" was given again exclusively as a patent to the Swedish inventors von Platen and Munters on September 26, 1922 due to a patent which was granted to the company Elektrolux in connection with von Platen and Munters' *gas cycle* driven by molecular weight differences—which was a 'markedly wrong decision.'
The company Siemens-Schuckert-Werke protested vehemently against this granting of the patent and initially prevailed in four courts of lower instance. The German supreme court ruled however, thereafter, in a "doubtful" verdict—which was probably due to the unexpected death of Reichsgerichtsanwalt [attorney at law] Mittelstädt three days before the hearing—definitely against the company Siemens-Schuckert-Werke. This paved the way for the company Elektrolux. The cooling aggregate was manufactured and distributed in nearly unchanged design by the company Elektrolux until now.
In the United States the competitor did not dare the process and paid about 1 million marks [former German currency] to the company Siemens-Schuckert-Werke. In those times it was a great amount of money but in the light of decades of production and world-wide marketing eventually a small one.

[65] Geppert[162] thought of atmospheric air as carrier gas. The molecular weight of NH_3 (17) and air (29) do not differ considerable so that a cycle of the carrier gas based on this would also have been an option (see Section 7), but Geppert ignored that. The diffusion coefficient plays naturally also a significant role. By using Lewis's model context between heat transfer and mass transfer, mass transfer and heat transfer could be optimized simultaneously. Dannies, H.[57] (1950) investigated within the framework of the theory of diffusion machines the mass transfer in diffusion refrigeration machines.

height h' into the absorber via the indicated metal trickle sheets and returns from there via an opening, beginning at the level of the solution sump[66], into the inner tube of the solution heat exchanger and in it back again, then later into the thermosyphon pump 8.

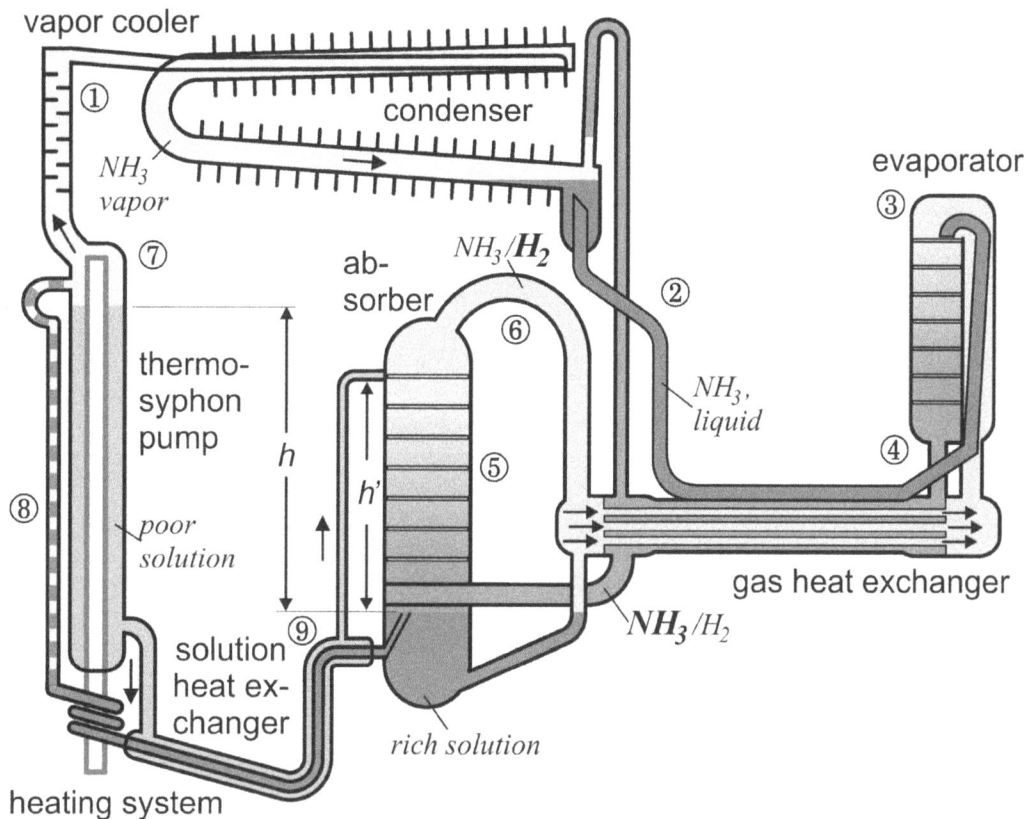

Fig. 6.5: Electrolux cryothermal apparatus according to VON PLATEN *and* MUNTERS

- The vapor from the generator passes a slightly downward running tube and gets to the end of the *condenser*.
- The liquid refrigerant (condensate) which is collected at the lowest point of the condenser flows through *line 2* to the upper end of the inlet pipe of the evaporator where it evaporates (diffuses) due to heat extraction into the poor refrigerant gaseous mixture arriving from the absorber via the gas heat exchanger.
- On its way downward through the evaporator the liquid refrigerant comes in contact with the gaseous mixture already enriched with refrigerant which therefore undergoes a gradual increase of its partial pressure so that the vaporizing temperature increases toward downward.
- Due to the absorption of ammonia the gaseous mixture in the evaporator becomes heavier on average than in the absorber. According to higher gas flow rates the mean partial pressures approach to each other in the evaporator and absorber whereby the density of the circulating gas trends slowly to an upper threshold due to the flow resistance.

---

[66] whereby in the bottom of the absorber the solution sump remains constant

. From the absorber to the generator the height difference h' has to be overcome only statically which increases due to flow resistances up to the outlined value h. This can be easily done by aid of a thermosyphon pump by which the solution becomes lighter at No. 8 than at No. 9 and is swept away by the rising gas bubbles, respectively[67].

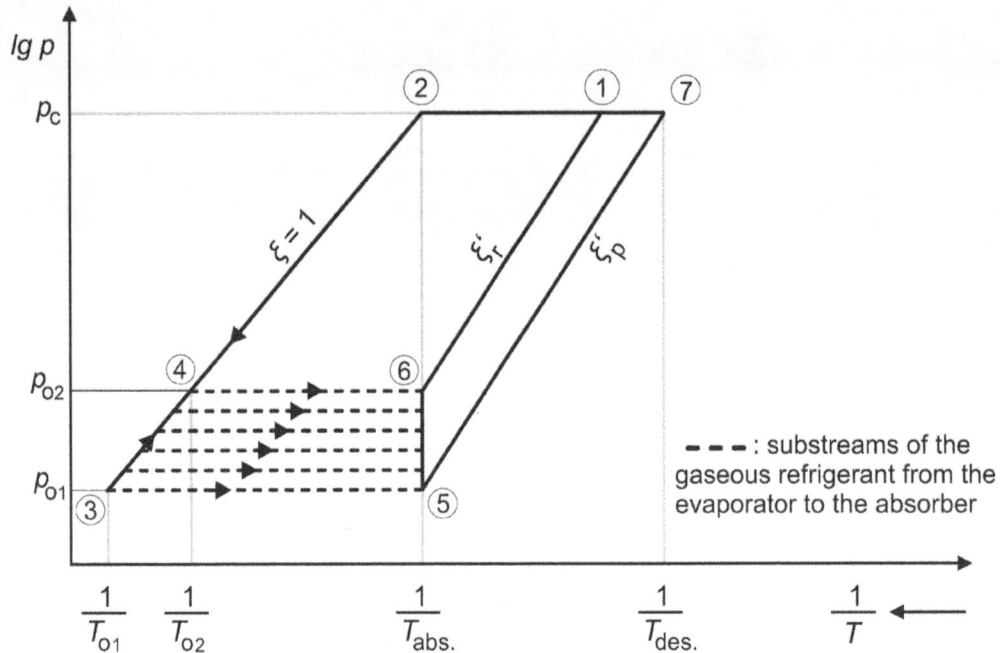

Fig. 6.6: Working process of the Elektrolux cryothermal apparatus (modified according to KRUSE—see Picture Credits, Section 14)

The working process of the cryothermal apparatus is shown in Figure 6.6[172]:

. Due to the refrigerant vaporization into a hydrogen atmosphere the cold production does not take place at a constant temperature but in a range of temperatures. The refrigerant entering the evaporator vaporizes initially at the pressure p_{o1} and at the evaporation temperature t_{o1}. With increasing vaporization the gaseous mixture enriches the refrigerant vapor much more so that the partial pressure of ammonia rises to p_{o2} and the corresponding vaporizing temperature up to t_{o1}.

. Point 3 describes the state of the refrigerant at the beginning of the evaporator, point 4 the state of the refrigerant at the end of it. The refrigerant is absorbed by the absorbent in the absorber so that its partial pressure drops from p_{o2} at point 6 to p_{o1} at point 5. In this state the refrigerant flows with the carrier gas back to the evaporator.

. The dashed arrow at the bottom directing from right to left (from absorber to evaporator) means the still remaining gaseous quantity of ammonia (partial pressure p_{o1}) in the ammonia-poor carrier gas when returning back from absorber to evaporator.

---

[67] cf. the solution transport according to the principle of the air-lift pump (mammoth pump) in BEHRINGER, H. (1930)[53].

The enrichment of the refrigerant in the absorption fluid is associated with the rising refrigerant partial pressure in the liquid from p_{o1} up to p_{o2} (5 → 6). The state of concentration of the rich solution is therefore fixed due to the pressure p_{o2} at the end of the absorption and the temperature $T_{absorption}$. Thereby, for example, it is an option—if desired—to conduct the process in a way that the absorption proceeds at a constant temperature whereas with absorption machines without carrier gas the working medium pressure is kept constant so that the absorption temperature $T_{absorption}$ decreases with increasing concentration of the solution.

6.4.4 The Atmospheric Resorption Cryothermal Apparatus According to Altenkirch

Working for the company Siemens-Schuckert-Werke Altenkirch was forced due to the patent situation to divert to the jet propulsion of the gas cycle with nitrogen as carrier gas. He was very familiar with the principle of jet propulsion due to his work on the steam jet refrigeration machine—see also his mentioned contribution[22] on this type of machines in Section 11.1.1 (Appendix). (Machines of this type were for instance successfully installed in canteens.)

This development ultimately had to be discontinued due to the initiation of the patent disputes and was again taken up only at the end of World War II in Zwickau[68] (Saxony).

After that, Altenkirch could establish his own laboratory at his residence in Neuenhagen near Berlin where he started his pioneering work on water extraction from the air, dehydration, and air-conditioning for deserts (see Section 9).

After his ELEKTROLUX patents had expired, Altenkirch turned again toward the cryothermal apparatus in Zwickau. He continued to pursue the idea of an atmosperic resorption cryothermal apparatus of which connecting lines/vessels should made of stonework, rubber, plastic, or the like; which opened up entirely new and diverse options for applications. Thereby, he considered air as carrier gas again and the *partial maintenance of the pressure difference due to liquid columns* in connection with his fluid cycle by using the molecular weight difference as principle of propulsion for the carrier gas cycle.

It was only after World War II in the years 1946 to 1949 that he pursued these ideas in a more concrete way when he was able to work in collaboration with the known refrigeration engineer DANNIES (see the working diagram in Figure 6.7 for this) within the framework of war reparations for the SMAD[69] in Zwickau.

In Figure 6.7 the pressure scale is inverted so that the low pressures lie at the top and the higher ones at the bottom of the picture, which corresponds to the actual altitudes of the devices. The partial maintenance of pressure due to liquid columns allows the reduction of the share of carrier gas from the pressure balance total and therewith the decrease of losses due to the imperfection of the unavoidable gas temperature exchanger by use of carrier gas.

---

[68] see Section 8

[69] Soviet Military Administration in Germany [translator's note].

According to the example in Figure 6.7 an entire maintenance of pressure due to liquid columns would have required an 8 m column height, thus, nearly two-third of the pressure difference is needed to be compensated by carrier gas up to a ceiling height of about 2.5 m.

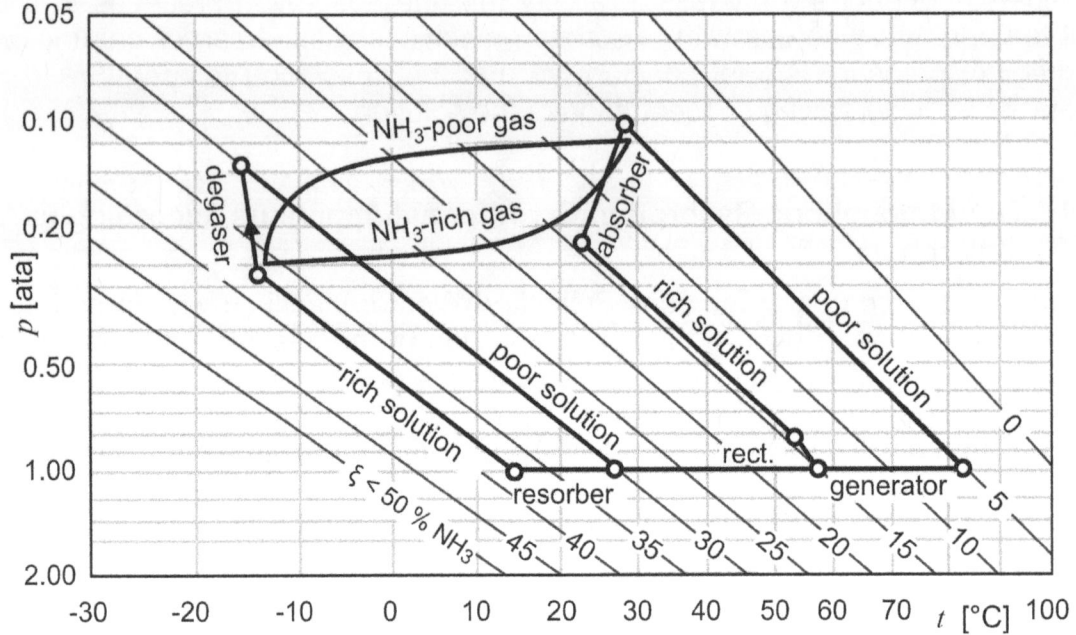

Fig. 6.7: Atmospheric cryothermal apparatus—working diagram

For the atmospheric resorption cryothermal apparatus and all motionless refrigeration machines (e. g. also the steam jet refrigeration machine) it should be noted that Altenkirch considered manufacturing devices made of only one material like glass, quartz, Cellon[70], stoneware, or porcelain which appeared to be particularly suitable[141] due to its chemical inertness and low thermal conductivity so that tube connections are unable to act as thermal bridges between cold and warm vessels.

6.5   The 'Therme'—The Inverse of the Cryothermal Apparatus

A (likewise motionless) inverse of the cryothermal apparatus Altenkirch called *Therme*. According to today's terminology *this term is identical to the term absorption heat pump*.

---

[70] The trade name for cellulose acetate manufactured by Deutsche Celluloid Fabrik, Eilenburg, Germany (source: http://en.wikipedia.org/wiki/Cellulose_acetate), retrieved 2013-08-02 [translator's note].

7 Reversible Air Treatment

In this patent specification[145] Altenkirch describes the novelty of his methods for air treatment as follows:

'All states of nature which are not equilibrium states can be utilized to produce energy, for example in shape of power or heat. One case of a lack of equilibrium, which is very frequently available, is the dryness of the atmospheric air. Therefore the dryness of air can be utilized for the production of energy, for example also for generating temperature differences, be it that one wants to reduce the available temperature for cooling purposes, or that one wants to increase the available temperature for heating. Thereby, a certain amount of water is evaporated which is lost in the atmosphere. By using the discovery, in those cases where there is only a limited amount of water available, and the temperature differences, which shall be generated, are not especially large, considering that, the more quantities of heat shall be generated or bound.

The invention can also be used to solve the opposite task, viz, to separate moisture from the atmospheric air. That means that the state of air should be further distant from the state of equilibrium, be it due to water shortage and therefore water is required to be extracted from the air, or be it because air is needed, which is drier than the available atmospheric air, for more difficult drying or cooling purposes. For such cases when air and moisture have to be separated, quantities of heat must be expended. Due to the invention the separation of air and moisture should even then be feasible when in fact sufficient quantities of heat are available but the utilizable temperature difference is relatively little such as with the temperature difference between irradiating solar heat and shadow.

Regarding these mentioned methods the atmospheric air undergoes a certain range of states between the method's initial and final humidity. The invention consists in the fact that the change of moisture in the air is driven beyond a threshold of this range of state and afterward is reversed. That can be carried out by aid of an appropriate absorbent which permits that the former change of state can be taken place within a limited range of temperature and that this change can be reversed again within another limited range of temperature. For this, a counter current between the air and the absorption solution will be purposeful when air and solution undergo strong changes of their partial pressure or their concentration during their circulation, respectively.

It is known per se that atmospheric air and absorption solutions can be brought into contact, be it that the air should be dried hereby, or that the absorption solution should emit absorbed quantities of water to the dry air again. A novelty regarding suchlike methods is, however, to drive the changes of humidity of the air temporarily beyond the threshold given by the initial and final state of the process.'

From 1930 onward Altenkirch devoted all his abilities to the challenges of water extraction under extreme conditions, allowing greenhouses to be operated on the Steppe, and to utilizing solar power for air conditioning.

At the beginning of these pieces of work the studying of conventional methods led Altenkirch to the assessment that it was hitherto not yet possible to find an easy way to accomplish the desired conversion of humid air nearly reversible and with a minimum of expenditure in energy. Only with changes of the state of the air which proceeded along the line of saturation had viable ways become known.

Fundamental for his research was the knowledge that partial pressure changes of the water vapor in the air which were caused by material exchange/mass transfer with contacting media (e. g. solid absorbents, like wood, silica gel, but also liquid surfaces like brine)—as well as with the sliding temperature changes of the Lorenz process—are only then reversible when the hygroscopic media which absorb and emit the vapor show the same partial pressure gradient / (slope of chemical potential) in flow direction as the water vapor so that the driving partial pressure difference for the material exchange/mass transfer can remain everywhere nearly constant and small.

For diverse tasks regarding the transformation of the chemical potential (say, the partial pressure) of the water vapor in the air or in hygroscopic substances, there are related analogies to corresponding tasks of heat transformation for which—as already above mentioned—in both cases the thermodynamic laws offer every option.

So one will find similarities, that are in the following section described, with Altenkirch's developed solutions regarding multistage connections, in the driving part of absorption machines, but here with pressure compensation by carrier gas (in this case with air) and water as refrigerant, as well as with extension of this type of machine to periodic absorption aggregates.

As driving energy for the material/heat transformation Altenkirch chose, in accordance with his interest in simplicity and saving resources, the natural potential difference of the temperature differences between the heat from the sun and the cooling in the shadow.

A prototypical procedure is the air dehumidification inside a periodic absorption aggregate toggled by the position of the sun and with silica gel as a sorbent. The water vapor (in the air) is the working medium and flows into a cycle which is opened to the exterior—a so-called open circuit—in which, for instance, the dry external air is transmitted into a room for drying products, from which it returns to the atmospheric air again after a water uptake, etc.

The temperature difference between sun and shadow is transformed into differences of the chemical potential. Vice versa, differences of humidity can be transformed into cooling/heating.

A typical device developed by Altenkirch is shown in Figure 7.2. It consists of two canals which are alternately heated or cooled (e. g. automatically toggled by the position of the sun) in which the humidity load of the hygroscopic substances increases in a vertical direction. While one canal is heated, the other is cooled. The canals are connected with each other at the top or the bottom according to the process conduct. In the warmer canal the air flows upward due to the thermal drive, in the other downward.

Analogously to the internal heat exchange by inserting regenerators, recuperators, and the like, into the thermodynamic cyclic processes, Altenkirch here used similar methods for the realization of the internal mass transfer. To this end, he noted[27]:

'Regarding the common cyclic processes used in the refrigeration technology, viz, those which are driven by external work and those driven by external heat, often an internal heat exchange is used which connects the single stages of the process with each other. This counter current heat exchange can be realized both with recuperators and regenerators. Stirling and Ericson should be mentioned here and the insertion of the regenerator into cooling air machines which promoted the knowledge of the thermodynamic interrelationships so extraordinarily. Of great importance was also the internal heat exchange regarding the method of air liquefaction according to *Linde* whereby a temperature difference of around 200 degrees was bridged by a heat exchanger with which—especially in the start-up phase—a multiple of those heat quantities was transformed which were bound by the throttling process. Likewise a multiple of absorption heat could be exchanged in the solution heat exchanger inside the absorption machine.

In these processes of internal heat exchange, the external work is completely inferior. The similar quantity of substance undergoes a process, which leads successively to a lower temperature downward and then upward again. In the borderline case very large amounts of heat can be exchanged while only very slight amounts of heat are required at the reverse points for the maintenance of the process.

But not less interesting is the fact that there are also thermodynamic processes similar in kind possible in which an internal material exchange connects single states of the process with each other. If, for example, a gaseous mixture flows successively in counter current through two pipes which are separated by a semi-permeable wall, and enables the change of the composition of the gaseous mixture at the reverse point, a material exchange by diffusion occurs and the gaseous mixture undergoes a process in which it is conducted through an area of higher or lower load of one or the other component and back again. Thus, for example, a relative humid quantity of air can be conducted through a state of great dryness by aid of such a counter current diffusion apparatus separated with a vapor permeable wall whereby only a small fraction of the amount of substance which is exchanged in the diffusion apparatus is taken from the reverse point.'

In the working diagrams 7.1, 7.3, and 7.5 the change of state on the vertical lines is meant to be an idealized isotherm mass exchange with the hygroscopic substances in the canals.

The changes of state of the humid air along the outlined straight saturation lines of the hygroscopic substances occur in an equilibrium (after several switch-overs) with the thermally layered hygroscopic substances in the canals or are caused by thawing-out/moistening at the saturation line of water. They are indicated with dashed lines, the changes of state of hydrogen with dotted lines. The circulating direction of the air is indicated with arrows, the heat exchange is emphasized by vertical, the mass exchange by horizontal double arrows.

7.1 Reversible Humidification of Air (Water Extraction from the Air)

An example of a process conduct in Figure 7.1 is shown in a log p 1/t diagram for silica gel with the mass percents of water in gel as curve parameters. In this case of application the change of state done at the reverse point is the cooling of the air flow according to the change of state from p_4 to p_3 and t_{hot} to t_{cold} (in Figure 7.1). The content of humidity (partial pressure) of air drops hereby. Principally, the process is only possible due to:

(a) thawing out of water at the saturation line of water, or

(b) absorption of water vapor by hygroscopic material,

whereby in both cases the heat flows of absorption or condensation, respectively, are to be discharged simultaneously.

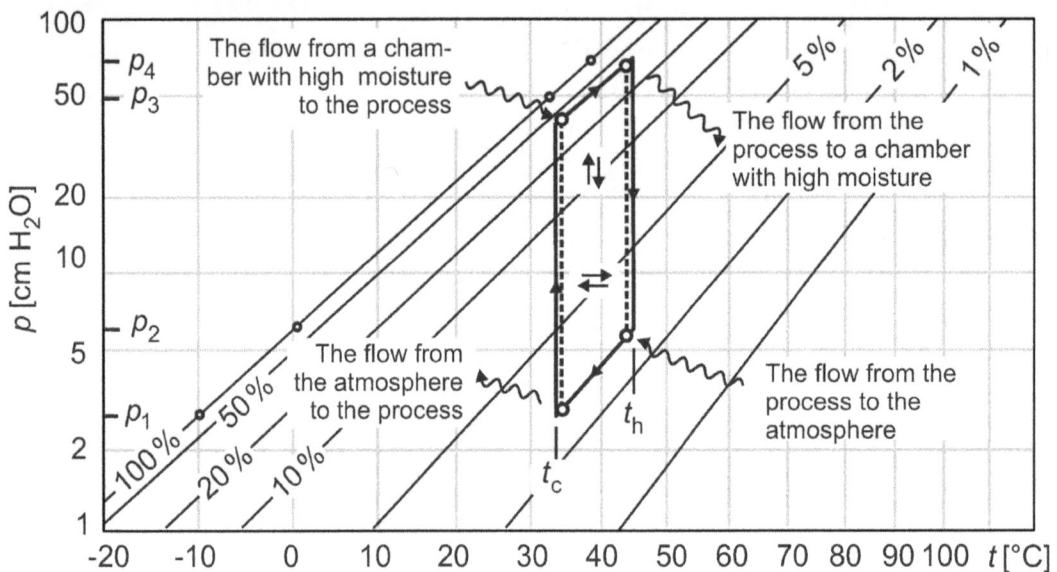

Fig. 7.1: Reversible air humidification (water extraction from the air)

In case (a) the purpose of the process is water extraction from external air which has a low content of humidity (desert climate). The further drying of the air until the partial pressure p_1 is a correlate not used here. Due to the continued transition of the hygroscopic materials (alternate heating/cooling of the canals due to, for instance, the changing position of the sun) the achievement of the saturation line is enforced.

In case (b) the purpose is the air-conditioning to keep the products moist.

In the exemplary cases of a technical implementation according to the Figures 7.2, ... , 7.4 the warmer canal has the temperature t_{hot} = 45 °C, the cooler the temperature t_{cold} = 35° C.

The air enters—in the case of Figure 7.1 with a partial pressure p_2 from underneath into the warm canal with the temperature t_{hot} and leaves after enrichment with water from the hygroscopic substances with the higher vapor pressure p_4.

The now moistened air can reach—immediately after some cold/warm switchovers of the canals—a sufficient content of humidity by taking up water from the hygroscopic filling materials of the canals so that it can be led at the location y into a room which has to be kept moist—in the exemplary case at temperature 35 °C.

From there it returns with a somewhat lower partial pressure p_3 after water emission to the goods that have to be kept moist—in Figure 7.2 a condenser is inserted between the canals, which enables an extraction of water in the middle of the overall height of the apparatus due to reaching the saturation line.

The air, leaving with the lower partial pressure p_3, passes the colder canal and emits more humidity to the hygroscopic substances situated here, thus it returns in a dried state to the atmosphere due to the necessary water extraction up to the partial pressure p_1.

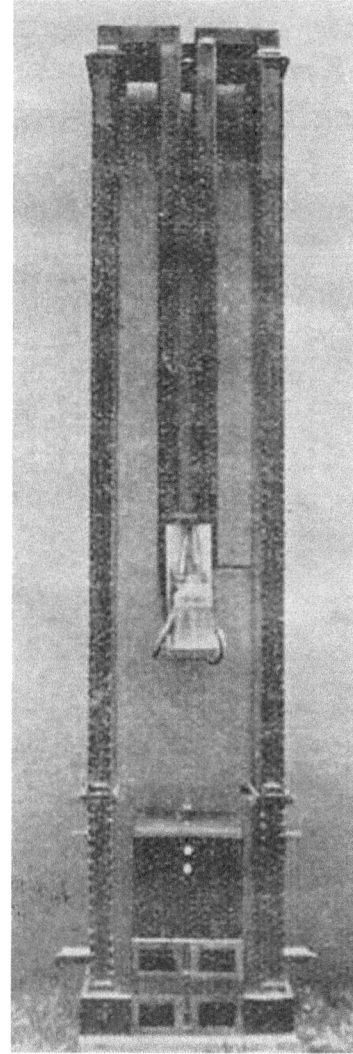

Principally, the hygroscopic substances could also proceed in the direction of the (straight) arrows as a mechanical cycle. In this exemplary case the "transition" of the hygroscopic substances is caused by the heating or cooling switch-overs (horizontal and vertical double arrows) like in a periodic absorption machine.

Due to the presupposed change between warming and cooling due to the position of the sun—or due to an other controlled heating/cooling—and the therewith caused circulation due to the gravity, the flow direction reverses automatically so that it takes place a complete exchange of the two canals with their fillings—and their regeneration—like in a periodic absorption machine. (There the gas stream is also reversed due to the alternated heating.)

Due to the temperature difference in both canals the increase of the water vapor partial pressure of the air is reached after several switch-overs at the reverse point from p_2 up to p_4 (see Section 6.4)—and thus the load of moisture of the there stored hygroscopic masses is increased—until the dew-point lies above t_{cold} and the saturation line of air as in Figure 8 at 40° is reached. After the successive cooling down of the air to $t_{cold} = 35\ °C$, water precipitates in a condenser inserted at the reverse point —or it takes place the desired humidification/moisturization of the room and its stored material, respectively, due to the gaseous mass transfer—without traversing the liquid aggregate state!

Fig. 7.2: Apparatus for air humidification (water extraction from the air)

In the following graphic presentations of the process conducts, the dashed lines represent the changes of state of the humid air due to heat and mass transfer to/from the hygroscopic materials. The spatial development of the state of the touched hygroscopic materials themselves is indicated by solid lines. The arrows at the air streams show the direction of the movement of the air, the arrows at the solid lines the change of state of the hygroscopic materials with fictitious motions in counter current toward the air.

7.2 Reversible Air Desiccation—Space Cooling

The air to be dried enters (Figure 7.3) with the partial pressure p_3 into the cold canal with the environmental temperature t_e. Whereas in the previous Section 7.1 a condenser had been inserted between the canals at the reverse point of the air flow, here an evaporator at the low end is situated between the air outlet from the cold canal at temperature t_e and the air inlet into the warm canal at temperature t_{hot}.

The air from the degasser/evaporator (cooling/drying room) here can absorb humidity under heat extraction at t_e (perhaps by inserting a counter-current heat exchanger) and thus causes cooling. Thereby, the temperature t_e must be above the dew-point for the partial pressure p_2.

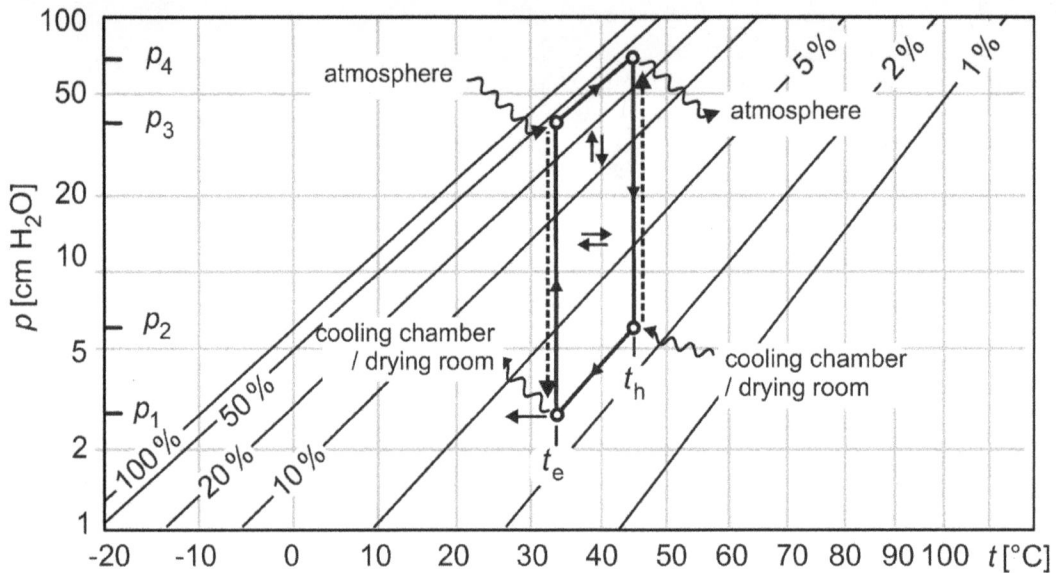

Fig. 7.3: Working diagram of reversible drying of air (space cooling)

The air comes from the cooling/drying room with the partial pressure p_2 and goes over into the warm canal from which it leaves with the partial pressure p_3 and returns to the atmosphere again.

7.3 Reversible Air Desiccation (Augmentation of Cooling)

It is very interesting to note the easily resulting opportunity for the augmentation of cooling as follows: As outlined in Figure 7.5 the atmospheric air with the temperature t_e and the water vapor pressure at p_4 is fed into the warm canal and there pre-dried up to p_2. The air is then at the state of the intersection point with a horizontal dashed line originating from point (p_2, t_e) and the saturation line for pure water with the resulting dew-point of circa 7.5 °C.

Shown in this case of application, the change of state to be executed at the reverse point proceeds from point (p_2, t_e) to point (p_1, t_{cold}) with a constant water content in counter-current to the temperature layering of the hygroscopic substances (p'_2, p'_1) which are regenerated with high temperature—this desic-cation can also be executed with ice by inserting a heat ex-changer $(p_2, p_1.)$ Afterward, the air is able to take up water from the local hygroscopic materials while flowing successively through the cold canal (upward) and along with it augments the expended cooling in the ratio $(p_3 - p_1) / (p_2 - p_1)$ which was needed for cooling of the air from t_e down to t_{cold}. This increase is, as expected from thermodynamics, greater the smaller the temperature difference $t_e - t_{cold}$ is.

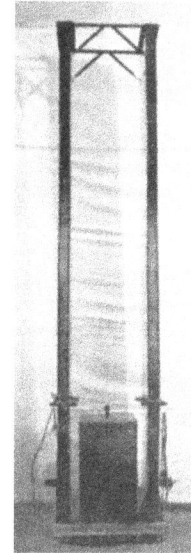

Fig. 7.4: Reversible desiccation (experimental apparatus)

In the exemplary case this change of state at the reverse point, as mentioned, is due to the dehumidification from other hygroscopic materials $(p'_2 - p'_1)$ which were re-generated $(p_6 - p_5)$ at higher temperature t_h with simultaneous cooling (from t_e to t_{cold}) due to the counter-current heat exchange from the air from p_3, t_{cold} to p_4, t_e (vertical oriented double arrows) that is closing the cycle—the coupling with the lower cycle is indicated through the semi-circular arrow.

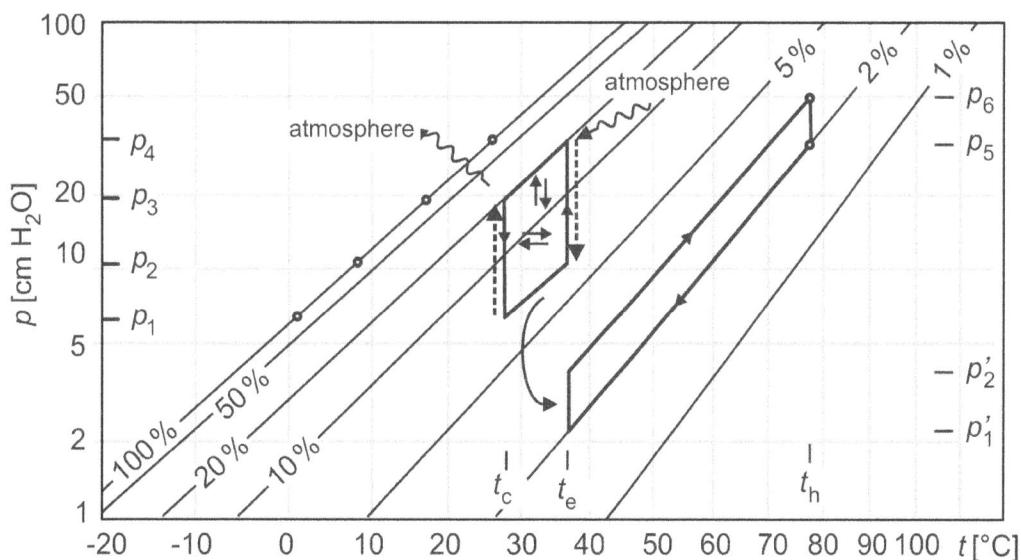

Fig. 7.5: Working diagram: Reversible desiccation (augmentation of cold)

7.4 Analogy with Multistage Designs of Absorption Machines

When comparing Altenkirch's latter methods outlined in the Figures 7.1 - 7.5 with his methods using multistage designs in the driving part such as according to Figure 5.15 with two-stage designs, it then shows that the discrete (stepwise) arrangement there by introducing carrier gas (air) becomes a continuous process whereby the generator vapor's intermediate absorption in the resorber is equivalent to the uptake of the water vapor in the left canal from the downward flowing air by the solid absorption materials and in the right canal the desorbed working medium is temporarily stored in the air until emission into the left canal.

In Figure 5.14, the working medium is fed from the low pressure generator at an intermediate stage through a vapor line to the resorber and is there absorbed. Further stages with higher and lower pressure level can be provided.

Here, the desorbed working medium is stored in the carrier gas, which enriches with working medium (humidity), before it is fed again in downward directed flow to the corresponding sorbents.

Here the aspiration pressure p_0 from the evaporator in Figure 5.20 is equivalent to the lowest partial pressure p_1 of the dried air leaving the left canal. The vapor pressure p generated in the high-pressure desorber in Figure 5.20 of the working medium vapor traveling into the condenser is equivalent to the maximal humidity loading of the air with the partial pressure of water p_4.

Fig. 7.6: Tropical dwelling with solar cooling

The condensing out of the working medium by leaving the generator of the upper stage of the discrete multistage design is equivalent to the thawing-out of the working medium (water) with reaching the highest working medium load of the carrier gas when cooling down before entering the cooler downward-positioned canal.

The thermal compressor produces a pressure difference total of $p_4 - p_1$.

7.5 Response to Altenkirch's Works on Air Treatment

In this context[75],[43] J. Schwarz's paper[109] should also be mentioned, with the title "The Heat *and* the Cold" and with the remark that there is nowadays a wide range of literature on applications of zeolite in the absorption heat technology.

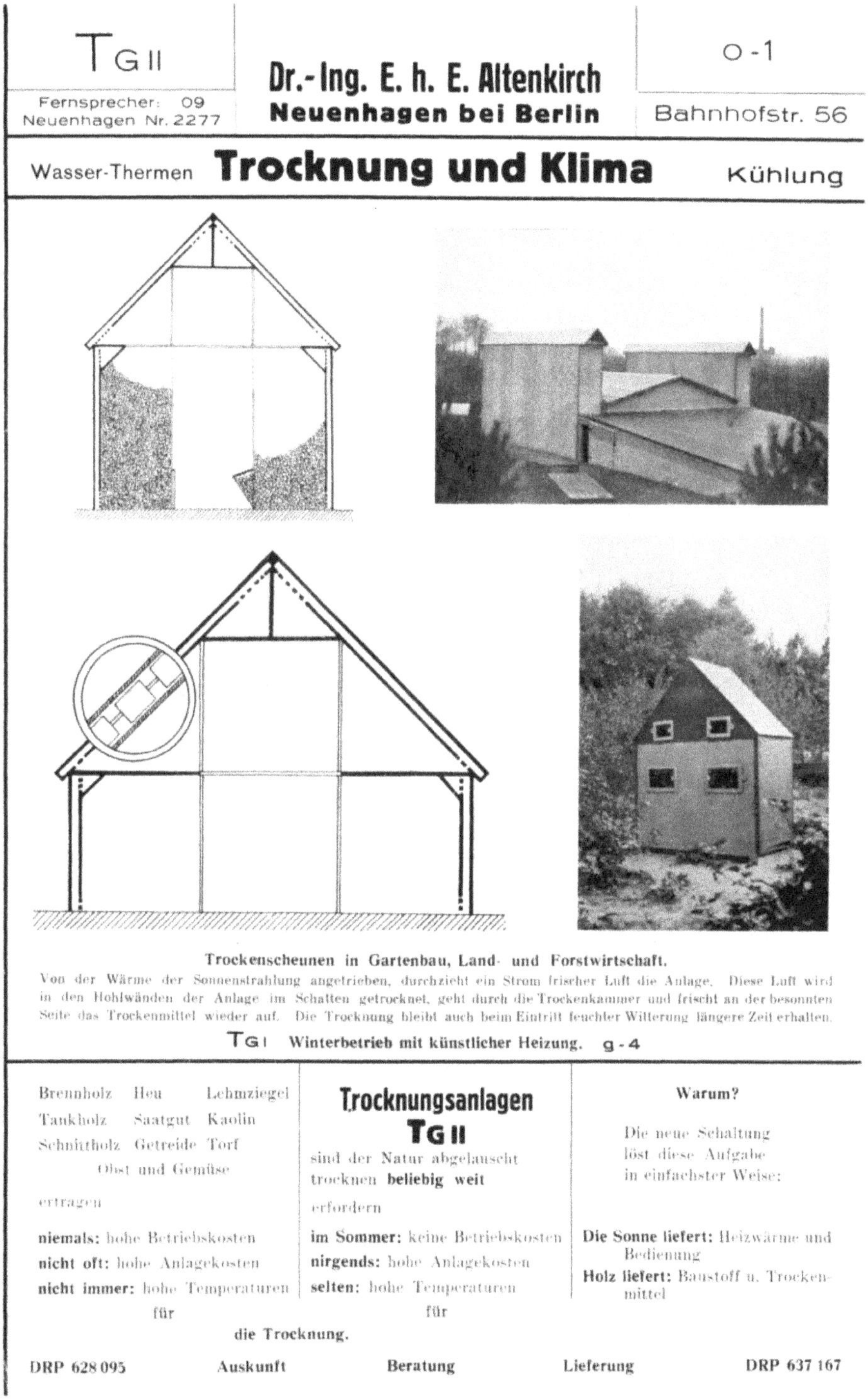

Fig. 7.7: Altenkirch's original leaflet

On the theme of cold production with solid absorbents the abstract of this paper is cited, 'A periodical working system of solid heat pumps with ecologically unobjectionable working materials like water and zeolite provides heat at 60... 90 °C and cold at 0...10 °C due to the possible high temperature lift. Thereby, the used working energy due to this double advantage is utilized very efficiently.'

Altenkirch himself built up an experimental plant on his estate in Neuenhagen for the demonstration of these facts (see Figure 7.6.) A leaflet from Altenkirch's production facility in Neuenhagen near Berlin is shown in Figure 7.7.

As the following excerpt (summary) from YAZICILAROGLU, S.[123] verifies, Altenkirch's method for the investigated application of water extraction was assessed positively compared with the state of the art at that time,

'An overview is given on the technical and economical options of water extraction from the air in arid countries and the energy demand for water production by air compression is calculated due to both a compression and absorption refrigeration machine. For certain circumstances results an optimum, but it does not allow yet to show prospects for solving the problem economically.

In comparison, Altenkirch's suggested absorption method provides the opportunity to get away with a simple construction and low operational costs and thus approaches a solution to the economic objectives.'

In the same investigation the problem of the technical suitability of solid sorbents is discussed critically, especially for wood as hygroscopic material. Altenkirch also made remarks in connection with this (ibidem, [27]):

'The first probational apparatus for house cooling without any moving parts is certainly not yet perfect. But it shows obviously the direction where thermodynamics is seeking to pursue new ways for cooling dwellings in the tropics which are, in the end, cheaper and more comfortable than the heating in winter at our latitudes. For this, with long switching periods of the device, quite large quantities of hygroscopic materials are necessary. Meanwhile highly suitable materials which are very cheap were successfully found.

It is self-evident that apparatuses developed for industrial applications and spoiled demands are more complicated. They cannot function without fans and automatic systems, and the artificial heating is in many cases essential. But the basic principle of it is mostly obviously recognizable at the simplest forms of them.'

For the dimensioning of constructions for the allowance of the simultaneous mass and heat transfer (between them exist model similarities according to the RAOULT's law; the first author) Altenkirch determined in [40] the parameters for the technical interpretation due to the formulation of cost criteria in consideration of the flow resistance and the laminar and turbulent start-up phenomena for the heat transfer and flow resistance. Altenkirch compared the turbulent and laminar flow regarding to this also in the start-up area of the process. The *fraction value length* of a slot, until which the temperature difference between air and wall has dropped to a fraction value (e. g. the half value), he used as a measure for the heat transfer. Altenkirch's conclusion culminated in,

'After simultaneously decreasing the speed and reducing the slot width, the pressure loss and simultaneously the demand in cooling surfaces along with the required space can be reduced extraordinarily far below that as yet achieved level, if it is successful to manage the practical, constructive, and manufacturing difficulties, which are connected with a reduction of the slot width. To these belong the necessity of a filter, the spatial arrangement, and the assurance of an even slot width, etc.

It seems that these difficulties would be managed most likely with regenerators for closed hot and cold air machines whereby the continuous regenerator according to Altenkirch[40],[154] provides additional decisive advantages. It should be remembered that under the name "aerodynamic turbine" nowadays a hot air machine has been developed for which a utilization of heat beyond the diesel motor has been tasked, which has a heat exchanger as one of its most essential elements, but which can still be largely perfected in the light of these considerations (cf. Z. Ges. Kälte-Ind. 50 (1943), p. 94).'

It should still be noted that the Second World War abruptly ended Altenkirch's continuation of these developments.

8 Work After the Second World War

8.1 Development of a Mobile Cold Storage System

'At the end of the year 1945, I accepted a contract from the Russian side to write a theoretical chapter for an elaboration of the container traffic. I was given permission to publish this paper, "Der Aktionsradius ortsbeweglicher Kältespeicher"[35] [The range of action of a mobile cold storage system].' It dealt with the development of a *mobile cold storage system* which should have the shape of a *cuboid*.[45]

Tab. 8.1: Definitions

a,b,c		edge lengths of the cuboid	[m]
k_{ex}		experimentally determined thermal transmittance value of the cuboid	[kcal h^{-1} K^{-1}]
r_{out}, r_{in}		outer, inner radius of the equivalent hollow spherical	[m]
δ	$= r_{out} - r_{in}$	insulation thickness of the equivalent hollow spherical	[m]
r_{aq}	$= 4\,\pi\left(\dfrac{1}{\alpha_a\,(r_i+\delta)^2}+\dfrac{1}{\lambda}\,(\dfrac{1}{r_i}-\dfrac{1}{r_i+\delta})+\dfrac{1}{\alpha_i\,r_i^2}\right)^{-1}$		[kcal m^{-2} h^{-1} K^{-1}]
$r_{aq.H}$		determination by evaluation of $k_{hollow\ spherical} = k_{ex}$ as to r	[m]
V_i	$= \dfrac{4\,\pi\cdot r_i^{\,3}}{3}$	volume of the inner spherical	[m³]
Φ	$= \dfrac{V_i}{a\cdot b\cdot c}$	space reductor	[-]
η		packing density	[-]
α_{out}, α_{in}		coefficient of heat transfer: environment → outer spherical, inner spherical → storage space	[kcal m^{-2} h K^{-1}]
V		volume of the storage space	[m³]
G	$= V\cdot\Phi\cdot\eta$	net chilled goods volume	[m^{-3}]
c_v		volume-specific heat capacity	[kcal m³ K^{-1}]
C	$= c\cdot G$	heat capacity of the chilled goods	[kcal K^{-1}]
λ		thermal conductivity of the insulating material	[kcal m^{-1}h^{-1}]
t_a		ambient air temperature	[°C]
$t_{initial}$		initial temperature of the chilled goods	[°C]
t_{final}		the still allowed final temperature of chilled goods	[°C]

In order to be able to talk about several important results, terms must be defined, which are indispensable for the understanding: Thus, Altenkirch introduced the term *equivalent hollow spherical* in order to consider the doors and heat bridges only once at the beginning. This hollow spherical has, by definition, with the same insulation thickness $\delta = r_{out} - r_{in}$ the same thermal transmittance value k_{ex} like the cuboid; k_{ex} can be easily determined due to its measurement: The spherical radius increases due to doors and heat bridges. The storage volume of the hollow spherical exceeds that of the cuboid. G is the chilled goods's volume to be accommodated there per m³ with the packing density η. The following results are to be stated:

8.1.1 Coefficients of Heat Transfer

Altenkirch defined the outer and inner coefficients of heat transfer with $\alpha_{out} = 10$, and $\alpha_{in} = 5$. For the warming-up of the chilled goods with known values for r_{out} and δ as well as the coefficient of thermal conductivity λ of the insulating material, the heat transmittance value k of the equivalent hollow spherical is determined according to the used formula (see Table 8.1). Letting C be the constituted water equivalent of the storage goods, the following time constant is decisive

$$\tau = \frac{k_{ex}}{C} \qquad \text{(Eq. 8.1)}$$

Therewith, the time is determined for the rise of temperature from t_a to a *still allowed temperature* t_{final} at an outside temperature $t_{initial}$ due to

$$z_e = \tau \cdot \ln \frac{t_{initial} - t_a}{t_{initial} - t_{final}} \qquad \text{(Eq. 8.2)}$$

8.1.2 Effect of Insulation

Borderline cases: With $r_{in}/r_{out} = 0$, the hollow sphere contains nothing else than insulating material. The storage capacity C is therefore = 0, likewise the storage time. With $r_{in}/r_{out} = 1$ the insulation thickness is = 0, and the incident heat is solely constituted by the external heat transfer with $\alpha_{out} = 10$. The k value is finite, but no small so that likewise a finite but *short* storage time results.

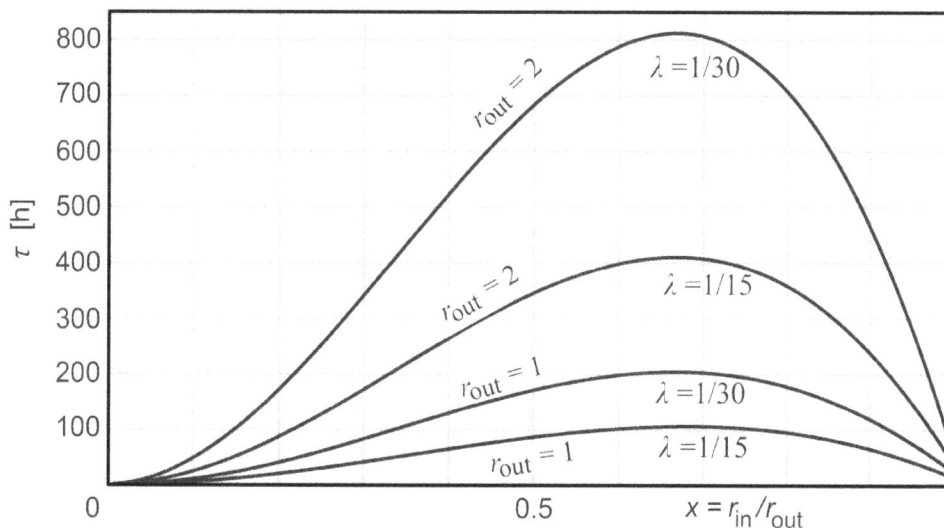

Fig. 8.1: Storage time dependent on r_{in}/r_{out}

Critical cases: Between these borderline cases lies a maximum of storage time. Interesting is their dependence on the quality (λ) of the insulating material with firm diameter of the spherical r_{out} = 1 m, as is shown in Tab. 8.2.

It becomes evident that with good insulation material (λ = little, consecutive numbers 1-3) the best insulation thickness (r_{max}) changes with the quality of insulation (λ) only little but the effect of insulation z_{max} itself however largely.

With bad insulation material (λ is larger; consecutive numbers 4-7), the best insulation δ (from 0.3 to 0) changes largely with the quality of insulation (λ), the effect of insulation (z_{max}) however little.

Until No. 3, the insulation thickness for good insulation materials results in

$$\delta_{r\,max} > 0.3 \cdot \rho \qquad \text{(Eq. 8.3)}$$

and is hence too big in practice. Thus, Altenkirch comes to the statement,

'Namely, it is less important to achieve the ultimate storage time z_e, but rather to transport the largest amount of the chilled goods in a constituted time with due safety. For example, let it be assumed that the arithmetic product of the chilled goods's volume and the elapsed time until reaching the still permitted final temperature z_e shall be as large as possible.' To this, his consideration, Altenkirch referred due to the maximization of the product $G \cdot z_e$ (which he called "cold transport capacity"). For good insulating material (λ little), the solution[71] results then in

Tab. 8.2: $\rho = 1$ m
Relative storage times and cold transport capacities dependent on the insulation thickness.

Nr.	λ	r_{max}	$\delta = 1 - r_{max}$	z_{max}
1	0.03	0.67	0.33	170
2	0.06	0.68	0.32	87.5
3	0.15	0.69	0.31	38.5
4	0.3	0.72	0.28	22.1
5	0.6	0.77	0.23	14.2
6	1.5	0.91	0.09	10.2
7	2.0	1.0	0	10

$$\delta_{G\,z_e\,max} = \rho/6 = 0.5 \cdot \delta_{r\,max} \qquad \text{(Eq. 8.4)}$$

He summarized: 'Thus, the best utilization of the cold storage system is achieved in an insulation which is half as large as the value obtained for the longest storage time. From the derived formulae the huge importance of a minimum thermal conductivity of insulation which cannot be compensated by an increased insulation thickness can be seen.'

The ratio of the cold transport capacity $G \cdot z_e$ for a constituted insulation thickness δ to the cold transport capacity for $\delta=0$ (curve with the parameter "$G \cdot z_e$") as well as in the same way the reduced maximum storage time z_e (curve with the parameter "z_e") dependent on $x = r_{in}/r_{out}$ is plotted in Figure 8.2 whereby the insulation thickness results in $r_{out} = \rho \cdot (1-x)$.

---

[71] In the original the term "$r - \rho = \varphi/6$" was written falsely, it is being replaced by "$\rho - r = \rho/6$".

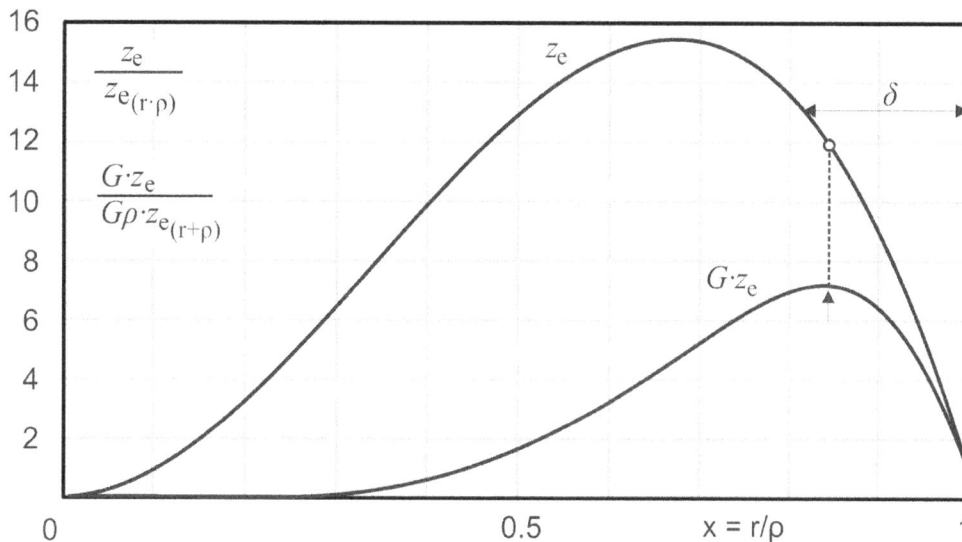

Fig. 8.2: Relative storage times and cold transport capacities dependent on the insulation thickness

8.2   The Extension of the Usable Temperature Range[72]

The extension of the temperature range to a variety of technical tasks by utilizing the following resources was Altenkirch's interest:

· The potential of work (exergy) of coal, measurable with the high equilibrium temperature of approximately 3,000 °C of the coal combustion with the system coal/oxygen. Exploiting it, should be feasible due to the heat transformation at high temperatures and had prompted him to start with intensive studies in his later Zwickau[73] time. The conservation of natural resources and purification of the atmosphere motivated him thereby early on.

· The waste heat of steam power plants

· The outdoor cold to increase the yield of the cogeneration

· The solar energy, especially for water extraction from the air in the steppe (greenhouse in the desert), and for air-conditioning

Altenkirch considered the extension of the temperature range of the steam engine for low and high temperatures due to the insertion of heat transformers with which the no longer usable temperature ranges by the steam engine should be developed due to the augmentation of heat.

The limits of exploitability due to the steam engine were constituted toward low temperatures due to the no longer manageable large vapor volumes of the water, toward high temperatures due to the critical phenomena of water, and due to the limitations of material stress caused by too high pressures and/or temperatures, respectively. The results of this work had not been published—they shall therefore be covered here:

---

[72] It deals with Altenkirch's first theoretical considerations about pairs of substances, equilibria, heat flows, and process cycles, apart from concrete technological objectives.

[73] Zwickau, a Saxony town in Germany, known for its coal mining district until the late 1970s, and the birthplace of the classical romantic composer Robert Schumann (1810 – 1856). [translator's note].

In particular, options for the upward extension of the usable temperature range of the steam engine were investigated. Realistic temperatures on the steam engine were taken as a basis with a condensation temperature at $T_C = 42$ °C and a generator temperature at $T_G = 295$ °C.

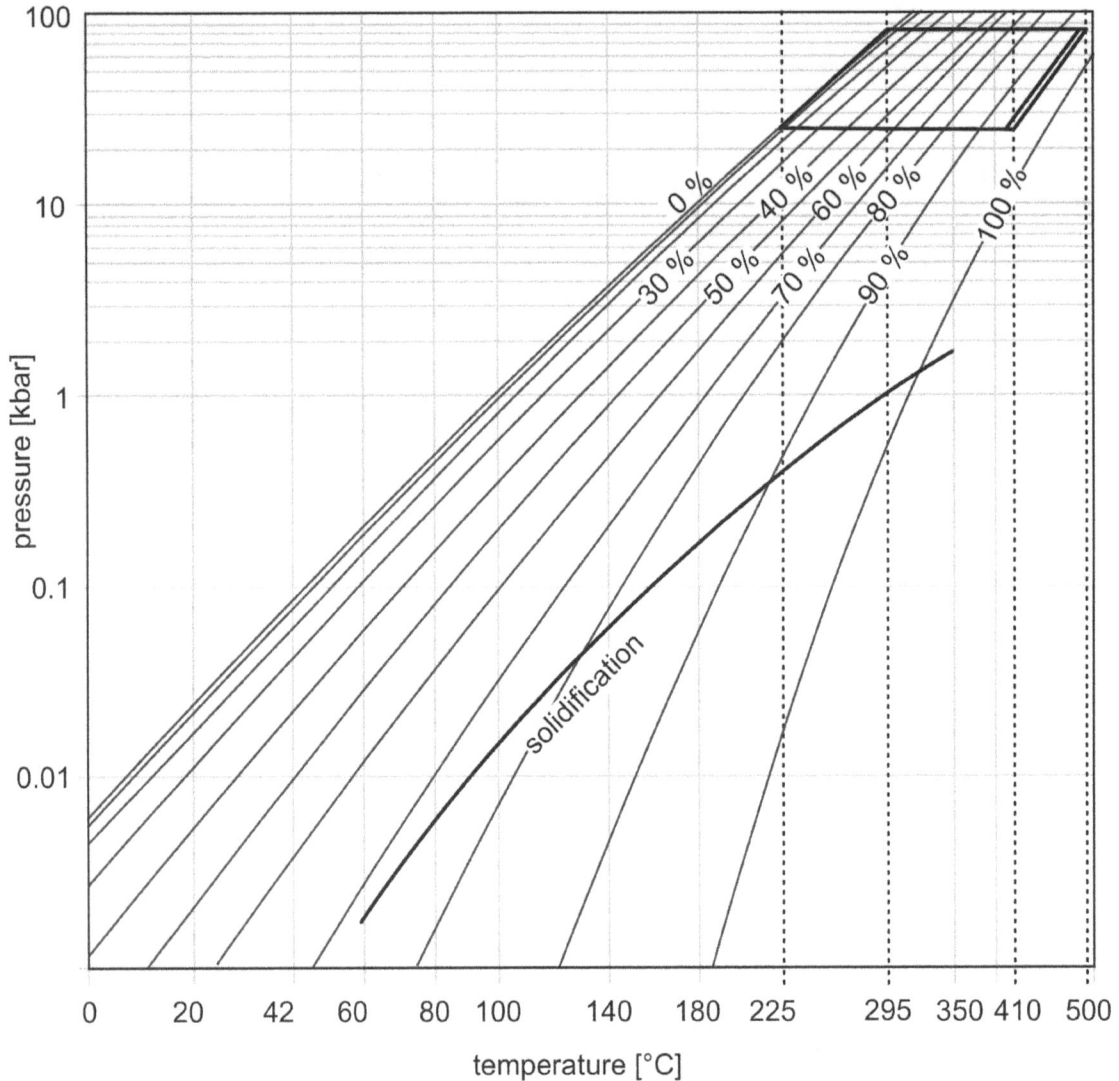

Fig. 8.3 : Solution plot for sodium hydroxide water

Without an upstream heat transformation the theoretical yield of work W from the thermal heat T_G fed to the steam engine at Q_G would result in

$$\frac{W}{Q_G} = \frac{T_G - T_C}{T_G} = \frac{\Theta_C - \Theta_G}{\Theta_C} = 44.4\ \% \qquad \text{(Eq. 8.5)}$$

The heat transformer should consist of a cascade of two absorption heat pumps, both powered by the working medium water: The *first* had been thought to be an interconnection of two at high temperature working vapor-coupled periodic machines. In the driving part of this interconnection this chemical reaction should take place:

Ba(OH)$_2$ $\Leftrightarrow$ BaO + H$_2$O (with bridging a temperature difference of

t_G [Ba(OH)$_2$]- t_A [Ba(OH)$_2$] = 1,000 °C - 510 °C = 490 °C),

and in the cooling part the reaction:

Ca(OH)$_2$ $\Leftrightarrow$ CaO + H$_2$O with a temperature lift of t_R [Ca(OH)$_2$] - t_V [Ca(OH)$_2$] = 500 °C - 300 °C = 200 K.

The pressures were inferred from tables of chemical equilibria with a *generator/resorber pressure at 1 kbar and an absorber/degasser pressure at 0.001 kbar* (the latter was envisaged to be equalized by carrier gas). Then, for the downstream second heat transformation the system NaOH-H$_2$O proved appropriate:

Tab. 8.3: State variables of the heat transformers

	T [°C]	p [kbar]	System	Heat [J/mol]	Con-Verting factor	Coupling
1	T_{GBaOH} =510	1.0	generator Ba(OH)$_2$ $\Rightarrow$ BaO+H$_2$O	-118	1	external thermal heat
2	T_{ABaOH} =510	1/1000	absorber BaO + H$_2$O $\Rightarrow$ Ba(OH)$_2$	118	1	
3	T_{RCaOH} =500	1.0	resorber CaO + H$_2$O $\Rightarrow$ Ca(OH)$_2$	127	1	
4	T_{ECaOH} =300	1/1000	degasser Ca(OH)$_2$ $\Rightarrow$ CaO+H$_2$O	-127	1	4 with 5
5	$T_{C\,NaOH}$ =300	84	condenser NaOH +H$_2$O machine	37	6.62	
6	T_{VNaOH} =225	26	evaporator NaOH +H$_2$O machine	-37	6.62	8 with 10
7	T_{GNaOH} = 500	84	generator NaOH +H$_2$O machine	-37	6.62	5 with 2 and 3
8	T_{ANaOH} >= 410	26	absorber NaOH +H$_2$O machine	37	6.62	
9	T_D=295	84	generator of the steam engine	-118		9 with 6
10	intermed. vapor 295...42	84...26	steam engine			

The solution plot of Figure 8.3 verifies that the generator according to line 7 can be inserted at 500 °C between the pressure levels 84 and 26 bar to absorb the heat of 118 kJ/mol according to line 2 from the Ba(OH)$_2$ absorber and the absolute value of heat listed in Table 8.3 from the resorber of the Ca(OH)$_2$ machine at T_R [Ca(OH)$_2$] = 510 °C.

The generator of the NaOH/H_2O machine receives thus a thermal heat at the mean temperature of 505 °C and at 84 bar (see Figure 8.3). Table 8.3[74] provides an overview over the entire system.

The converting factors guarantee the correct thermal balances of the couplings.

The Ba(OH)$_2$ $\Rightarrow$ H_2O reaction in the generator of the barium machine in No. 1 shall release water vapor with the mass of 1 mol which can be absorbed due to the reaction CaO + H_2O $\Rightarrow$ Ca(OH)$_2$ in the resorber of the calcium machine in No. 3. The process at 0.001 kbar proceeds in the opposite direction. From the resorber and absorber of this tandem process total, the heat quantity of 118 + 127 = 245 J would be released.

This, in No. 2 and 3 developed heat quantity total has to be absorbed by the generator of the NaOH machine in No. 7. This can solely be achieved when the circulating water (vapor) quantity in the NaOH machine is increased accordingly. The converting factor thus amounts to 245/37 = 6.62.

Moreover, the degasser of the calcium machine must receive the absolute value of heat of 127 J (column 4). There the condenser of the NaOH machine according to column 5 is used.

The condenser of the NaOH machine also releases the heat quantity of 6.62 · 37 = 245 J/mol due to the value of the converting factor. When 127 J/mol are subtracted, still 118 J/mol remain on the condenser of the NaOH machine. Together with the absorber of the NaOH machine of 245 J/mol, 118 + 245 = 363 J/mol are initially available from the NaOH machine for the heating of the steam engine.

However, still 245 J/mol from the intermediate vapor of the steam engine at 225 °C are to be supplied to the evaporator of the NaOH machine, thus, on the steam engine only the absolute value of heat of 118 J/mol at 295 °C will remain—as it ought to be due to the First Law of Thermodynamics.

Tab. 8.1 : Gain of work of a the steam engine by co-generation

Type of Machine	Kind of Internal Heat Transfer	Gain of Work [kWh]
Periodic Absorption Machine	Without Heat Recuperation	487
ditto	With Temperature Equalization	582
Continuous Absorption Machine	With Heat Exchanger	987

Then, the effect of the heat transformer exists "only" therein that the intermediate vapor had to be pumped up from 225 up to 295°C, which is equivalent to a yield of 245 · (295-225) / (297+273) = 30.1 J/mol. The steam engine performs without the heat transformer a work of 118 · (295-42) / (295+273) = 52.6 J/mol, which corresponds to an augmentation of the working efficiency by 57.2 %.

Explaining the facts, it should be noted that here inside the steam engine a purely isentropic pressure relaxation with the work output is presupposed so that

[74] The heat flow r in the aggregates was approximately calculated according to Clausius-Clapeyron with the gas constant R = 8.31 J K^{-1} mol^{-1} which resulted in $r = R \cdot \lg (p/p_0) / (T^1 - (T_0)^{-1})$. Thereby the pressure ratios are for columns 1 and 2 = 1000, likewise for column 3 and 4; for 5 and 7 and 6 and 8 = 84/26. The temperatures were plotted according to the table.

the taking-out of the intermediate vapor provides a reversible opportunity to adjust the temperature of the usable enthalpy of the vapor to the evaporator temperature of the NaOH machine for heating the evaporator.

Realistic calculations were carried out with periodic and continuous absorption machines as well, inter alia by taking into account the material data, the heat transfer losses, and container losses. The gain of work which can be achieved with coupling a steam engine which performs normally 10,000 kW, with different heat transformers is shown in Tab. 8.1

Of course, a decisive enlargement of gain of work would be still achievable by developing additional binary blends for the heat transformation beyond the temperature range reachable with sodium hydroxide water. Thus, in the exemplary case, a temperature increase from 300 °C up to 1,000 °C would enlarge the theoretical gain of work up to 6,700 kW.

The resource conservation due to improved utilization of the coal as raw material was Altenkirch's lifelong great interest. The considerations, reflected here in summary, are submitted records of Siegfried Unger (the first author of this book), made during the collaboration as a research assistant on these projects.

Already in those times, Altenkirch saw resource conservation as a societal responsibility and considered avoiding the exploitation of fossil fuels from early on. He even envisaged the usage of renewable energy resources like geothermal heat and outdoor cold for space heating (see the next section). The principle of outdoor cold for space heating, which he had designed at that time, had not been put into practice yet.

Even more interesting connections with larger effects were developed, envisaged due to the use of binary systems with metal alloys, which are not discussed here in any greater detail.

8.3 Further Work

Analog connections with ammonia/water absorption machines as heat transformers were also developed for the extension of the usable temperature range of power generation toward low temperatures, that means below 42 °C where the conventional steam turbine must fail due to the rapidly growing specific volume of the water vapor.

The developments for *climate control in cold rooms* were also started in Zwickau and later finalized. Altenkirch summarized the results, after returning to his Neuenhagen research establishment, in book form (to this end read see Section 9.2.2).

Design documents of an *atmospheric resorption cryothermal apparatus* working with ammonia/water for a capacity of 10,000 kcal/h ($\approx$ 11.6 kW) with air as carrier gas had been worked out in Zwickau. Thereby the predetermined limit of overall height (and thus the proportion of the pressure maintenance by liquid columns) prompted trade-offs between the size of the pressure equalization due to the carrier gas, and a further pressure decrease in the machine by reducing the average working fluid concentrations according to the principle of resorption. We are talking about the minimization of the overall losses which result from the sum of the losses due to the protraction of cold in the carrier gas cycle, and the insufficient rectification of the generator vapor.

9 Work at the Neuenhagen Research Establishment

9.1 Utilization of the Outdoor Cold

Moreover, he dealt with the utilization of outdoor cold for space heating, which inspired the West Berlin entrepreneur Dr. Bernhard Vogeler, Zehlendorf[75], for planning relevant projects. So, in [46] an article by Dr. Vogeler was published, entitled, "Heizung ohne Kohle durch Erdwärme und Außenkälte"[76]. Here, the absorption machine is used as a heat pump. Due to the refrigeration technology, the minor cooling-down of the groundwater or of earth layers (geothermy) is "pumped up" as an extracted heat of an average temperature level of ca. $t_1 = 10\,°C$ to a thermal heat of ca. $t_2 = 30°C$.

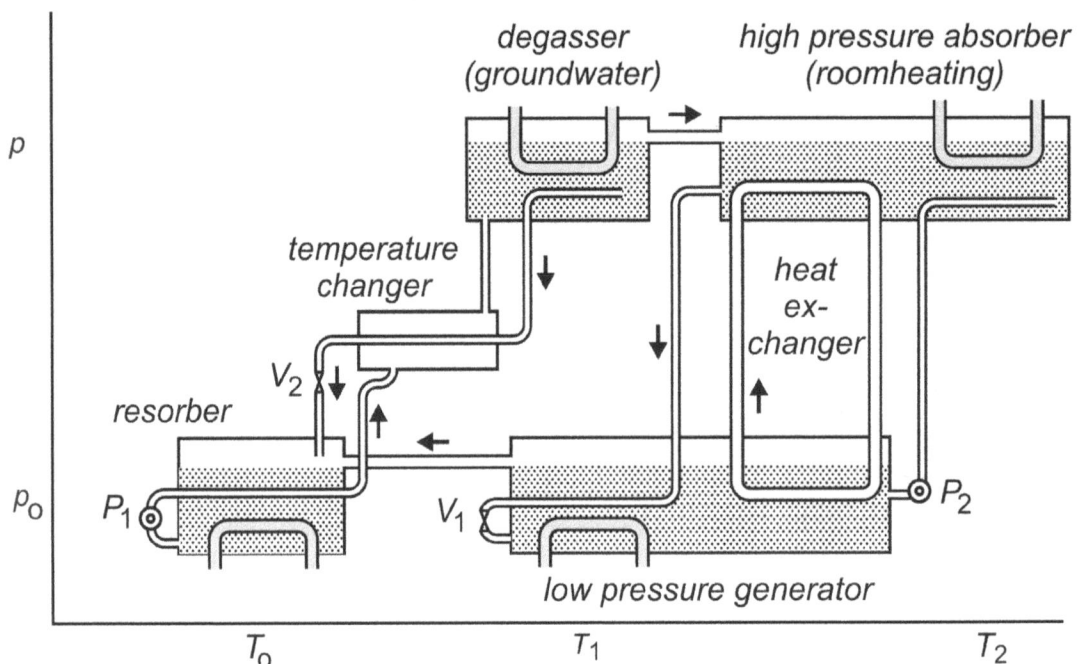

Fig. 9.1: Schematic of residential space heating by use of terrestrial heat and outdoor cold

The driving force for this process is the exergy due to the temperature difference between the heat reservoir at an average temperature T_1 and a reservoir (e. g. outdoor cold) at the lower temperature T_0 of ca. -10 °C. In this case the cyclic process of the absorption machine is conducted into the opposite direction as previously described: The condenser becomes a high-pressure evaporator (heated by groundwater), the absorber becomes a degasser (heated by groundwater as well), the evaporator becomes a condenser (cooled by the outdoor cold), the generator becomes a high-pressure absorber (which delivers the thermal heat for the space heating).

To achieve the temperature T_2 Altenkirch applied his method of overlapping tem-

---

[75] A Berlin district [translator's note].

[76] "Heating without coal due to geothermy and outdoor cold" [translator's note].

peratures (cf. Section 5.5)—in this exemplary case we are talking about the overlap between the high-pressure absorber and the low-pressure desorber.

Due to his interest in simplicity and low-maintenance Altenkirch had chosen the cryothermal apparatus as a cold-generating system. For the limitation of the column heights—as required in buildings—Altenkirch used the principle of resorption, combined with a slight pressure equalization due to air as carrier gas—as an environmentally friendly (green) and preferable foreign medium, free of maintenance, compared to other media.

From the pressure equalization due to liquid columns, the spatial lowering of the vessels of higher pressure is shown, that means the *high-pressure absorber* which delivers the thermal heat, and the *high-pressure evaporator* which is heated by the groundwater—the latter was chosen due to the applied principle of resorption instead of a *high-pressure evaporator*.

The realization of temperature bands in the resorption cycle enables also the improved utilization of the heat transfer media (in terms of the Lorenz Process—as noted several times earlier).

An article, found in Altenkirch's archive, on utilization of the outdoor cold for the apartment block heating was used for the schematic of Figure 9.1, reconstructed according to Alternkirch's original intention.[155]

The "transmitter"[77] causes the heat exchange for the "overlapping" of the temperatures of the high-pressure absorber and low-pressure generator. With this, the required initial flow temperature for the space heating was achieved at the expense of a slightly increased need of groundwater.

9.2 Publications in Book Form

9.2.1 Absorption Refrigeration Machines 1954

This work could only be brought to print after his death (unfinished) by his last collaborators, H. Voigt, and the first author of this report, Siegfried Unger—the chapter "Kryothermen" (cryothermal apparatuses) is absent (this chapter is only stated in the introduction of his work in page 11, paragraph 3, from below).

The contents of this book as to continuous processes are explained and illustrated at length in Section 5 "Absorption Machine as a Heat Transformer" in the actual report—the work's missing part is treated in Section 6 "Cryothermal Apparatus —A Motionless Heat transformer."

9.2.2 Climate Control in Cold Storage Rooms

The work contains comprehensive and concrete fundamentals, partly in the form of an optimal regulating function, and in the form of all required substance values, immediately usable as tables or diagrams. The movement of the state of the air from an actual to a set value of state in consideration of the time constants of all regulating variables (the cooling of the evaporator surfaces, the moisturization on moist surfaces of the chilled goods, and the like) was carried out by taking the "shortest

---

[77] It corresponds to a heat exchanger: That means a liquid medium which increases its temperature by taking up heat, without the need for boiling. In general it is a solution of polyalcohols, or a salt-water solution [translator's note in collaboration with the first author].

route." The state of the air is outlined in an i-ξ diagram. Altenkirch was enabled by his former work in the field of air treatment to carry out profound investigations on the climate control in cold storage rooms. They led to the development of a number of basic methods and means for the targeted manipulation of the state of the air.

The procedure focuses on the primary influence of the state of the air due to the actuators "cooling," (c), "heating," (h), and "moisturization," (m).

The primary dehumidification due to regenerative sorbents (e. g. with a wet air cooler powered by brine) were excluded by Altenkirch; to this end he commented:

'The regulation of humidity in cold storage rooms due to a special dehumidification system as a fourth actuator might also cause the removal of water vapor at cold storage room temperatures. To this end the absorption heat is evolving which leads to a change of state of the air in line with an equal enthalpy in the direction of higher temperature with lower water content and causes decrease of the relative humidity which is too strong and reversed due to the cooling-down to the desired amount of the cold storage room temperature. The regeneration of the absorption medium might occur outside the cold storage room due to the supplied heat. A decrease of the refrigerating capacity results then therein that the refrigeration machine can be more or less relieved from the condensation of water vapor, that means that it is at the same time no longer bound to the very low evaporator temperature which is required with a strong dryness, and therefore requires a reduced energy consumption for the driving force. For cold storage rooms in which a very low relative air humidity is required, a special dehumidification system with a practical benefit shall be provided.

An extension of the investigations in this direction would have however had little importance for the practically more important field of high humidity. For a relative humidity of 60 % an evaporator temperature is already sufficient which falls below the room temperature (0 °C) to -6 °C so that the absolute value of the additional heating (e. g. at -12 °C evaporator temperature) seems still bearable when the amount of humidity is not too high and the outdoor temperature as well as the heat supply from the outside is not too low, respectively.

For smaller facilities or transient operation with lower air humidity, the complication with an additional particular dehumidification facility would not be justifiable. The investigation in this direction is therefore deferred here, although it might be instructive to provide an overview of the whole field of humidity control in cold storage rooms without interruption.

For the following it is explicitly stated that the regulation of humidity shall exceptionally ensue due to the three actuators cooling, moisturization, and heating; a particular dehumidification facility remains outside the debate.'

The state of the air is described due to the variable pair (vector) $p=(t, \varphi)$ (t: temperature, φ: relative humidity), and is compared with the assigned set point vector p_r (t_r, φ_r) in the Mollier i-x diagram.

The chosen interference of the state of the air due to the actuators cooling, moisturization and heating leads as is generally known to the following changes of the states in the Mollier i-x diagram:

I: *Heating*, (h): Change of state at a constant water content x toward higher temperature and lower relative humidity.

II: *Cooling*, (c): In the exemplary case (Figure 9.2) the cooling takes place on moist and frosty cooling pipes at the temperature of -10°C and proceeds then in the *i-x* diagram on a straight line *i* = constant—starting from the actual point of state *p*, *t*—to the dew point (φ =1, t_o = -10 °C) (situated outside the outlined area at the right and lower half space of the saturation line running on the right margin). To this end both the temperature and the water content *x* of the air drops.

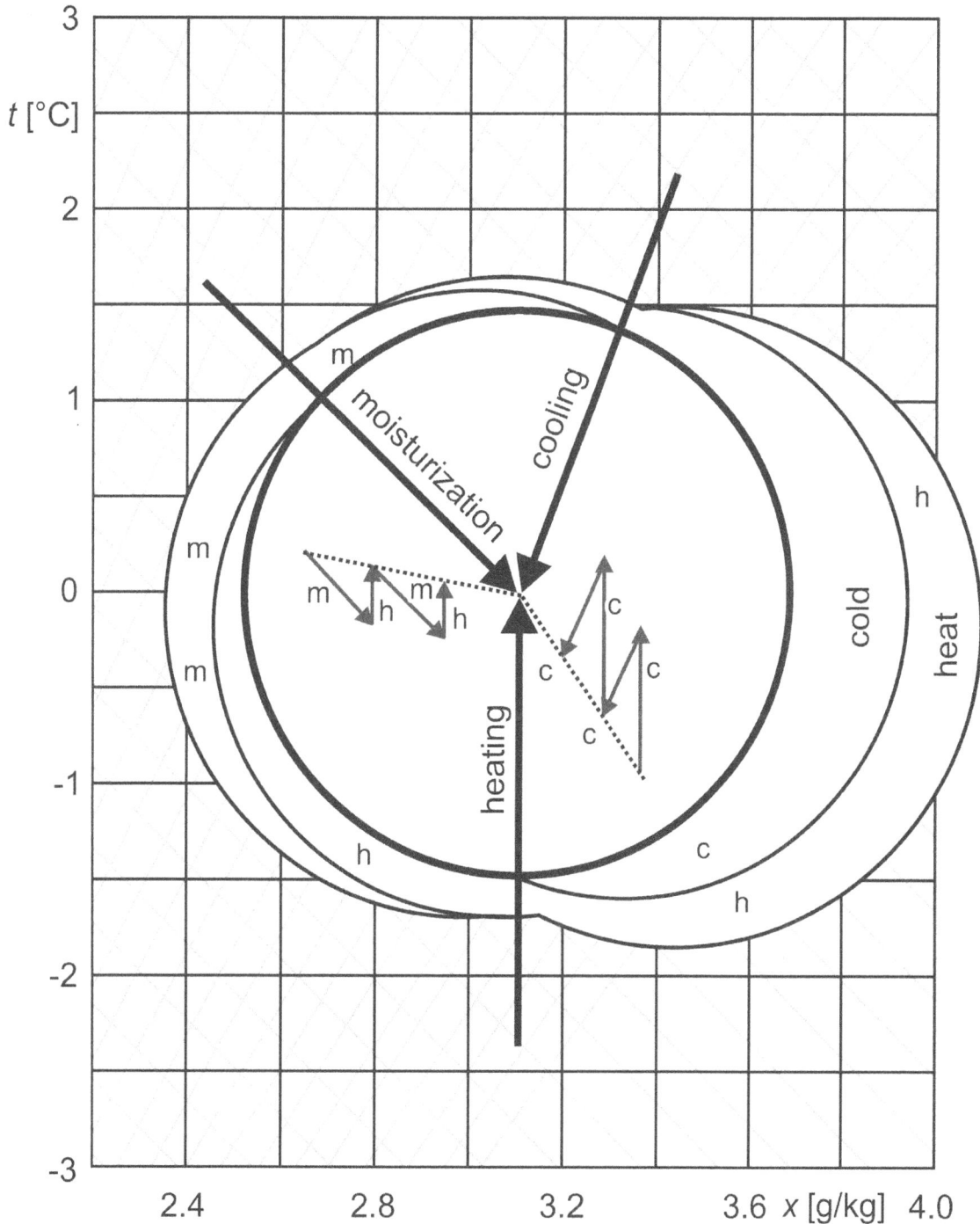

Fig. 9.2: Regulating diagram

III: *Moisturization*, (m): This change of state follows a line i = constant toward lower temperatures and higher content of humidity of the air.

For the deviation vector $\Delta p = p\text{-}p_R$ in Figure 9.2 there are 3 possible sectors which are limited due to 3 vectors V_I, V_{II}, V_{III}. This is outlined by solid lines with arrows directed inwardly to the center p_R whereby the arrows e_1, e_2, e_3 indicate the impact of the corresponding actuators, specified on the circumference in the regulating diagram, on the state of the air. Let $b_1 = \text{-}e_1$, $b_2 = \text{-}e_2$, $b_3 = \text{-}e_3$, the corresponding basis vectors (of a plain vector space) of which two of each are fanning out one of the 3 sectors. The sectors shall be fanned out as follows (see regulating diagram): Sector j due to the basis vectors $j+1$ *mod 3* and $j+2$ *mod 3*, $j=1,2,3$.

Active actuators: The actuators to be activated in a concrete regulating case of a state of the air, described due to the vector p, are determined as follows:

At first, the vector difference $\Delta p = p_R \text{-} p$ is established. For Δp both components are determined as to the basis vectors of each of the 3 sectors, that means the basis vectors, assigned to them. By definition, Δp covers that sector for which both components are positive.

The fastest shift of the vector Δp toward the center is then defined due to that base vector regarding of which the component of Δp is positive and maximal. This base vector then clearly determines the actuator to be activated, that means how it is plotted on the margin of the regulating circle with the designations h, c, m. In the regulating system these elements are stored and available, dependent on p and p_R.

Two examples show the iterative sequential zigzag movement of the point of the state toward p_R. Different modalities of regulation are discussed, inter alia a parallel regulation with a dead zone.

In the several sectors the following elements are active:
in Sector I – II : cooling + heating
in Sector II – III : cooling + moisturization
in Sector III – I : moisturization + heating.

For more information, see [43].

9.2.3 Fast-Running Regenerators (Publisher: VEB Verlag Technik 1952[40])

The realistic simulation of processes in regenerators due to the approximation of solutions of the underlying partial differential equations, which are only possible nowadays due to the deployment of a specific computer software, was achieved by Altenkirch in approximation using an iterative mathematical method[78]. His methodology is outlined in detail.

No major oversimplifications were made but the very conditions according to the types of flow (laminar, turbulent), inter alia in consideration of the Bernoulli acceleration and deceleration processes for the determination of the pressure drop and heat transfer, were concretely taken from comprehensive tabular overviews.

To this end, also a "general equation for the heat transfer in the pipe" is already used here, which was substantiated in detail at a later date [42].

---

[78] using the formulae and tables for the substance data which are to be plotted due to the dimensionless values *Re, Pr,* and *Pe'*

9.2.4   Delaying Function[41]

This regards the optimal float control for the "Compression Machine with Solution Cycle" (see upcoming Section 9.3.1).

9.3   Publications in Periodicals

9.3.1   The Compression Refrigeration Machine with Solution Cycle [37] 79

This is the title of a lecture, given at the conference of the German refrigeration association in 1948.

The achieved advantages of Altenkirch's proposals are:

- "The increase of an economic feasibility of compression refrigeration facilities due to the approximation of the external heat exchange processes to the Lorenz process whereby a saving on quantity and costs of the required heat-absorbing/-emitting media is achieved."

- "Moreover, with the concentrations of the working medium in the solution cycle, parameters are obtained, which allow to vary the pressure level in the machine. For the desired adjustments, the temperature ranges of the solution in the resorber, or also in the degasser must be brought into compliance with the required respective temperature ranges for the brine and cooling water; and the strength of the circulation of the solution cycle must be adjusted accordingly."

Moreover, these temperature ranges are to be changed temporally if necessary; also without alteration of the temperature ranges, it can be necessary to enhance or weaken the solution cycle, namely then, when, along with a power control of the compressor in line with the cooling demands, the aspirated quantity of ammonia is increased or decreased.

For the regular function of the machine it is important that the resorber is evenly loaded. To this end, the reservoir beneath the degasser must always be filled to prevent the solution pump from running dry.

This is, as is depicted in Figure 9.3, achieved with a regulating valve controlled by a float with stroke reduction for the introduction of the rich solution into the degasser. Further-

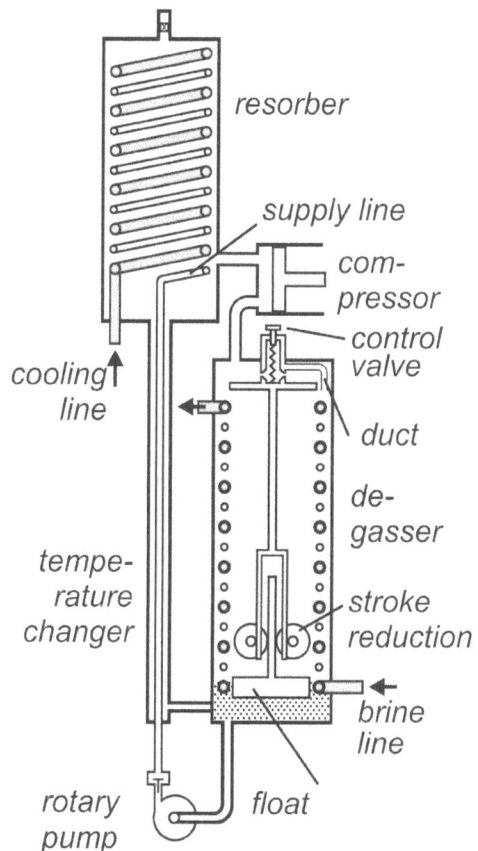

Fig. 9.3: Schematic of a compression machine with solution cycle

---

[79] According to H. Lotz's personal communication only one compression machine with solution cycle was executed since the droplet precipitation in front of the condenser was not yet adequately manageable.

more, in particular, the important prerouting of the solution in the resorber, and in the degasser are achieved due to the outlined heat contact through the wall of the pipe between both the external trickling, resorbing, and degassing solution, respectively, and in the pipe's counter-current supplied streaming solution.

A specific difficulty with the required self-regulation is thus the delay of the retroactive effect of the regulating impulse to the sensor, that means the lagging influence of the position of the regulating valve to the float. The increased or decreased introduction of the quantity of solution through the regulating valve on the degasser needs a certain amount of time (the "passing-through time") to be conveyed from the upper part of the degasser in a decreased quantity (because of the degasification) to the degasser sump, in order to influence the float which has to correct the position of the regulating valve. In the meantime, too much or too little solution necessarily enters the degasser.

Consequences are fluctuations of the fluid level in the float room due to insufficient dimensions, which can be enhanced in very unfavorable cases so strongly that the regulating assembly causes the machine's non-operability. That implies the task of clarifying the impact of delay dependent on the construction data.

In recent years, Voigt, H., *and* Eder, F. X.[121], *again took up Altenkirch's ideas and applied them to the Stirling process. A proposal emerged to extend the temperature band of heat exchange in the regenerator on the ends to the thermal coupling of the heating and cooling media whereby these advantages discussed here could become effective also in this process.*

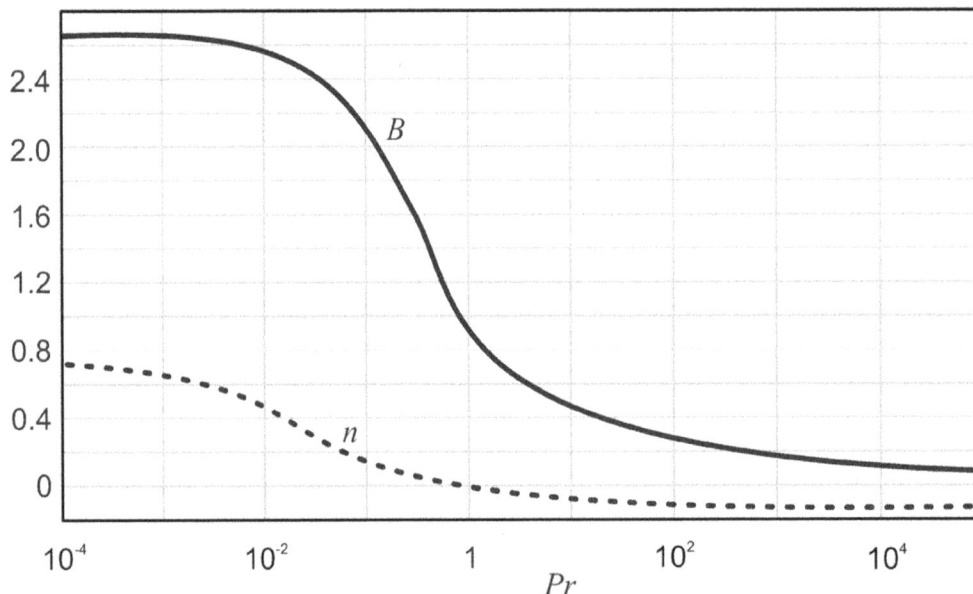

Abb. 9.4: General equation for the heat transfer in the pipe. The parameters B and n as functions of the Prandtl number

9.3.2 Delaying Function[41]

To control this impact, Altenkirch organized the "delaying function" in tabular form, which also should fall to an autonomous importance within the entirety of essential mathematical standard functions.

The principle of the Compression Machine with Solution Cycle was also taken up again in the more recent literature: Groll, E. A.[64].

9.3.3 General Equation for the Heat Transfer in the Pipe[42]

The equation which records the turbulent and laminar flow-type as well as the accompanying start-up phenomena, shall be rendered here. Its interrelationships shall be rendered completely and in summary here, so that the formula can be used. The steps (definitions) which are necessary to this end are the following:

- **The coefficient of heat transfer α_{lam} of the calmed laminar flow**
  with the pipe diameter of d is according to Nusselt

$$\alpha_{lam} = 4.06\ \lambda/d \tag{Eq. 9.1}$$

- **The consideration of the laminar start-up distance**

According to Hausen, H. (1950)[67] the correction factor for the consideration of the start-up distance is constituted by

$$\beta = 1 + \frac{0.0183 \cdot Pe \cdot d/L}{1 + 0.045 \cdot \left(Pe \cdot d/L\right)^{2/3}}\ {}^{80} \tag{Eq. 9.2}$$

Thus, we must write

$$\alpha_{lam\ start\text{-}up} = \beta \cdot \alpha_{lam} \tag{Eq. 9.3}$$

- **The coefficient of heat transfer α with the turbulent flow**
  With the turbulent flow the start-up distance is considered with the factor

$$\alpha = \Psi \cdot \alpha_{lam} \quad \text{mit} \quad \Psi = \frac{C \cdot Pr \cdot \Phi^{0.75}}{\left(1 + A\right) \cdot \Phi^{n}} \tag{Eqs. 9.4, 9.5}$$

wherein Pr, the Prandtl number, and the quantity Φ are interconnected with the Reynolds number Re and of which critical value $Re_{crit} = 2.320$ with

$$\Phi = Re/Re_{crit} \tag{Eq. 9.6}$$

the quantity A depends as well on Pr once more. The expression is shown as

$$A = \frac{4}{7} \cdot \left(1 + \frac{0.2166}{0.2888 + Pr}\right) \cdot \frac{Pr - 1}{\left(Pr + 1\right)^{1/6}} \tag{Eq. 9.7}$$

Eventually, the description for the exponent n is still required. It is shown as

---

[80] Pe (Peclét number) = d (pipe diameter, gab width, ...) · v (velocity) / (kinematic viscosity); Pr (number) = v (kinematic viscosity) / a (thermal diffusivity);
Re (Reynolds number) = v (velocity) · d (= pipe diameter) / v (= kinematic viscosity)

$$n = \frac{1 - Pr}{8 \cdot \left(Pr + \log\left(1.19 + \frac{2.96\, Pr}{1.32^{\log Pr}}\right) \right)} \tag{Eq. 9.8}$$

Through introduction of the variable

$B = \dfrac{(1+A)}{Pr}$ it can be written somewhat simpler $\Psi = \dfrac{3.25 \cdot \Phi^{0.75}}{B \cdot \Phi^n}$ (Eqs. 9.9, 9.10)

The coefficient of heat transfer α is then determinable dependent on Re due to the following decision scheme:

$Re < Re_{\text{crit}}$: $\quad \alpha = \alpha_{\text{lam}}$ $\quad$ (according to Equation 9.3)
$Re >= Re_{\text{crit}}$: $\quad \alpha = \Psi \cdot \alpha_{\text{lam}}$ $\quad$ (according to Equation 9.10)

with Ψ according to the Equations 3.13, 3.14, 3.15, and 9.10 as well as the description of the quantities B and n according to Figure 9.4.

Further publications are:
- *The impact of the finite temperature differences on the operation costs of compression refrigeration facilities with and without solution cycle* [38].
- *On the technical importance of the inverse Thomson-Joule effect in the hypercritical area, Altenkirch* [39].

9.4 Exposés: Planning of the Research Work for Authorities of the GDR[81]

9.4.1 Absorption refrigeration machines[82]

This exposé for starting-off the production of absorption machines in the former GDR contains a simultaneous overview authored by Altenkirch about all of his novelties introduced into the absorption refrigeration technology :

'"Introductory Notes.
Absorption refrigeration machines, as far as I know, are not manufactured in the GDR at the time.

Initially, it should be checked out if the commencement of the fabrication of these machines can be recommended.

Herewith, it has to be differentiated between the large absorption machines which are mostly pump based, and the small units working without pumps. The latter are called cryothermal apparatuses. They are used for the cooling of household refrigerators (e. g. gas refrigerator).

My working life is closely affiliated with both systems.

---

[81] German Democratic Republic: On the territory of the former Soviet occupation zone (today's eastern Germany), a communistic-oriented second German state existed from 1949-1990 [translator's note].

[82] Altenkirch named the collaborators involved in this research. These were: Dr. Niebergall (I), Maiuri (I and II), Dr. Nesselmann (II), Dr.-Ing. Dardin (II), Dr. Behringer (III), Dr. Dannies. (I, II, and III indicate the corresponding successive points.)

I. The Absorption Machines

Until 1911, the absorption machines were thought to be a rather imperfect type of machine, but despite that they were economically stable and competitive since they were driven by exhaust steam so that their operation was virtually free.

. In 1911, I applied for my patent, D.R.P. 278 076[160], in this I showed due to which measures the absorption machine can be developed to a nearly perfect machine. Due to papers in the periodical 'Zeitschrift für die gesamte Kälte-Industrie'[2] 1913 and 1914, I provided theoretical substantiations for the possibility of the enlargement of the heat ratio by 2 to 3 times, and far beyond 1, which had been previously thought impossible since the heat ratio and the quality grade were confused with each other. These works caused a considerable stir. They were not only reprinted in many countries but they also gave cause to the professors Merkel and Bosnjakovic in Dresden in 1929 to design new types of diagrams for binary mixtures, and namely at first for aqueous ammonia solutions with which the calculation of the new absorption machines could be carried out more easily."

This book was also translated into Russian and led to a refrigeration technology research station for absorption machines in Odessa, which was destroyed in World War II.

In Germany a series of large-scale refrigeration plants were built according to the new principle, especially for the chemical industry, for the gas cleaning at the long-distance grid gas supply, for the chocolate industry, and other purposes (A. Borsig, W. Niebergall).

In England the system was especially used for facilities to generate dry ice (Maiuri).

The consequences of this development are not yet by any means finalized, due mainly to the lack of sufficiently experienced experts. The commencement of the fabrication of the absorption machine in the GDR seems purposeful.

II. The Cryothermal Apparatus

This plan, which I pursued since 1911 to make also the pumps superfluous, led to the patent, DRP 395 421, from the year 1920, in this a continuous absorption machine without any pump was documented for the first time.

This machine was developed under my leadership as a refrigerator machine for the company Siemens-Schuckert-Werke (SSW). It had a very favorable heat ratio, also a low electricity consumption in electric heating, but it suffered from material difficulties with the applied absorptions solutions which came into question for temperatures below 0°C.

The result of this development was the successive patent, DRP 427 278 (1922), emanated by me, due to this the liquid cycle caused by vapor bubbles was shown on the cryothermal apparatus for the first time.

Two months later the Swedes v. Platen and Munters (ELEKTROLUX) applied for a absorption machine with pressure-equalizing carrier gas which was circulated by molecular weight differences. Although this machine had a much worse heat ratio (1/2 to 1/3), it could be operated with ammonia solution so that no material difficulties occurred and came on the market at an early stage (gas refrigerator in the United States, over 10 million in use).

It came to a patent dispute. The company SSW claimed the patent, DRP 427 278. The action for the annulment of the competitor was dismissed in both instances. Our infringement action was in both previous instances successful. In the last instance it was, however, dismissed despite a clearly favorable expert's report for us. (Our attorney in law passed away shortly before the appointment.)

As a result, the company Siemens-Schuckert-Werke sold the American patents to an American subsidiary company of the competitor. Nowadays air-conditioning systems for 15,000 und 25,000 kcal/h are manufactured serially and sold in large numbers (600 pieces/month) in the United States according to the principle of my patents, 395 421 and 427 278, that means without carrier gas with lithium bromide in water as absorption solution.

I was virtually taken out of the further development in Germany for the following reason: ELEKTROLUX used the earlier mentioned application, which was submitted two month after [my] patent DRP 427 278, to divert an application for a liquid cycle, and since this was failed due to [this] patent DRP 427 278, it was converted to an application which quite generally copyrighted the continuous absorption machines in which (including the gas cycle) all cycles are caused due to heat. This copyright was actually granted (despite the manifestly inadmissible extension) to ELEKTROLUX, even though I had applied for a machine (without gas cycle) which also exclusively was operated due to heat due to the patent, DRP 595 421, two years earlier, and even though I had applied for the sole liquid cycle, which was also run due to heat, two month before v. Platen-Munters's application which I already received as a copyright save that the copyright failed in the last instance.

This development had to be explained since its knowledge is indispensable for the understanding of the following descriptions.

In connection with the company Siemens-Schuckert-Werke I tried sustainably to bring about a consensus with ELEKTROLUX in order to prevent that our development experience did not get abandoned.

In advance we had covered all decisive ways of the further development with applications. This was pursued, for example, due to the patents: DRP 459 549, 500 301, 551 555, 653 357, and many others which emanated almost entirely by me. Unfortunately, however all attempts for the understanding with the objectives for a factual collaboration for the further development of the cryothermal apparatus were brusquely rejected due to the company ELEKTROLUX. This Swedish company could rely thereby on the previously mentioned heat patent whereas we did not receive the granted protection for the liquid cycle before the Supreme German court.

Since the patents for the further development still belonged to the SSW, a far-reaching further development of the v. Platen-Munters machine remained undone, which is barely changed in its relevant details since 1922; also after expiring of the SSW patents, ELEKTROLUX did not use them, probably for reasons of prestige.

After the predominant ELEKTROLUX patents had also expired, many companies (e. g. 18 alone in West Germany, 4 in Switzerland) focused on the reproduction of the ELEKTROLUX machine, apparently not considering improvements. The profit and the options for the export due to this factual situation could only fall to that company which took over the leadership for the further development and put substantial progress forward.

Only a very few engineers are well positioned for this task. I may include myself with them since I developed a special 'development procedure' for the cryothermal apparatus for the company SSW, in which ammonia absorption machines work at atmospheric pressure which, compiled from glass and rubber tubes in a few hours, can be operated below temperatures of 0°C. A rebuild for the testing of improvements ensued during the operation of the machines in a few minutes. (This was also demonstrated by me to a Russian expert in Zwickau in 1946.) The development time is thereby enormously shortened, also for complex dispositions. The thus modeled machines can afterward be manufactured from steel tubes.

Regarding my advanced age I must however limit myself thereupon, to instruct young capable people, to conduct this further development according to the given viewpoints.

I have already worked in this direction to be prepared. Upon the implemented development work for the occupying power[83], in three years from 1946 to 1948, I met the 25-year-old collaborator, Siegfried Unger[84], as an aspirant in engineering physics who is outstandingly talented and exceptionally interested in the eligible specialties: thermodynamics, fluid mechanics, and heat transfer. In 1949 I induced him to come to Berlin as my collaborator where he is now studying physics and mathematics at the Humboldt University. He also provided valuable assistance in the implementation of the mathematical research work: 15/18250[85] which was assigned to me in 1950.

Regarding a technically oriented development with the objectives for fabrication and export, a good natural engineer who is gifted in the art of construction and who also considers interests in production technology as well as a draftsman and an experienced laboratory mechanic would naturally be required. Surely such experts are partly known to me as well.

The company SSW withdrew entirely from the construction of refrigerators. They only sell products of other companies.

It should also be noted that I, in order to that my experience did not get abandoned, already in 1949 submitted a development request for the cryothermal apparatus at the Ministry of Planning of the GDR. It was defeated, since the economic efficiency would have not been proven.

Regarding the commencement of the development of absorption machines and the cryothermal apparatus, it is compelling that die absorption machines can be tackled only in connection with a plant of the iron industry. There is no precision work required but only welding work on sheets and tailored tubes. The pumps are obtained.

A predevelopment before the delivery of the goods is only in so far required as per se the known constructions for the manufacture have to be prepared. Regarding such large objects, innovations are not tested before in the companies but normally in connection with the new systems to be delivered. Serial manufacture is not relevant.'

---

[83] SMAD (Soviet Military Administration in Germany) [translator's note].

[84] The first author of the submitted report

[85] At that time, the internal designation inside the supporting company "Kältetechnik Zwickau"

9.4.2 Development of Compressors with Liquefiers and Evaporators for SO$_2$, CH$_3$Cl and Especially for Freon 12

'Refrigeration machines which are operating with sulfuric acid, methyl chloride, and freon 12 have been on the market for decades. A further development of these machines, which aims at a weight reduction with equal performance, is a task which all producing companies are continuously working on. This work also provided remarkable successes, which were achieved due to the increase of the piston speed by increasing the revs of the driving shaft of the compressor. Herewith one reached so high revs that a further increase encountered material and fabrication difficulties. Occasionally, even again the rotational speed had to be reduced.

Furthermore there are constituent problems which are broadly focused on, which are touching with the task closely. Thus, in 1951 the DKV[86] created a contest to critically compare the freon evaporators, which had become known, in order to make proposals for their new designs. To this end, especially the recirculation of the lubricant should receive attention.

The task specified in the headline can only be carried out within the framework of plants which manufacture refrigeration machines with the above mentioned refrigerants, which is for them a self-evident field. If standard models should be created, several companies had to be able to carry out the fabrication, so a close collaboration of the leading plants could be considered, in which an open exchange of experience with mutual criticism can lead to tangible considerations for the development of the standard types. However, a strict management of the process is required to prevent it from becoming too lengthy.

Initially, the standard types have to be determined, for which the constructions documents shall be created, but the further points of view must not be overlooked which lead to a satisfactory further development of this group of refrigeration machines either.

The weight reduction is not decisive. Decisive factors are the economic efficiency and operational safety. For instance, the more lightweight compressor can possibly be more expensive in manufacturing. The more lightweight condenser can give cause for a rise in operating expenses. The smaller evaporator can cause an inadequate holding of moisture. A rotational speed which is too high can lead to the operation being unsafe. Special attention needs to be paid to the molding difficulties for the freon compressors.

Also the best designs may not be possible to manufacture in a centralized location with a serialized fabrication system. This can be summarized as follows, the given task requires care that it is possible to find at least one operator who is up to it. When one is entrusted with the task, he must be given the opportunity to consult the manufacturing and operational experience of all plants which manufacture such machines. Otherwise a result could occur, which in the meantime is already being overtaken by a company working in parallel.

The task can be solved more easily when a plant, which has distinguished itself as especially good with construction, is entrusted with the task of development.

---

[86] German Society of Refrigeration and Air Conditioning, DKV
(source: http://www.dkv.org/index.php?id=115), retrieved 2013-02-12 [translator's note].

Since the company Kälte-Richter, Berlin, was excluded due to other reasons, the company DKK-Scharfenstein for instance would come into question.

The development costs quoted seemed too low, since, if we are talking about new constructions, they have to be tested beforehand.

9.4.3 On Freons

'As far as can be assessed, the procurement of freon 12, or other freons in the eastern zone[87] caused difficulties, but one cannot do well without freons as refrigerants, especially when the options for the export are considered. Freon machines are often already demanded although they may be not yet that necessary.

The non-toxicity, non-flammability, and odorlessness are indeed benefits which should not be underestimated, and technically the low pressure difference opposed to ammonia at an equal temperature difference of the refrigerating capacity is of a particular value since a one-stage operation at larger temperature differences in many cases is still possible whereas the ammonia machine already has to be disposed in multistage designs. Hence, the machines become partly simpler.

Furthermore, apart from this reduction of the pressure ratio, also the large vapor volume to be hoisted, which is required due to the low evaporation heat, meets the type of construction of the turbo-compressors which, with their rotational speed and their compact construction, will prospectively prevail in the course of time more and more over large-scale facilities, however there is an urgent need of smaller compression refrigeration facilities already today. Therefore, for instance, an efficient chemical plant, like the company Leuna-Werke[88], should be prompted to adopt a position for the option of freon production.

When the technical options for this are affirmed, it were to be determined by a survey among plants delivering refrigeration machines which need for freon is expected in the case of delivery possibility for the next few years.

To this end has to be noted that for freon an especially careful molding is required. Since currently different companies do not yet feel able to cope with these requirements, the real need can be slightly underestimated, but also the option of export has to be considered. Currently, freon is still rare and relatively expensive. Hence, the product exports are considered favorably. Since the generation is not easy and there are special presumptions of experience required, this price ratio will probably remain unchanged for many years. A displacement of freon by other refrigerants is not expected, in fact the reverse.'

---

[87] The so-called Soviet occupation zone on the territory of central Germany after the World War II up until the German Democratic Republic (GDR) as a German partial state with communist orientation was founded on October 7th, 1949 [translator's note].

[88] Originally founded as an ammonia plant in 1916, this chemical site at the place Leuna in Central Germany was and is producing a wide range of chemical products, such as methanol, mineral oils, and products based on brown coal processing as well as synthetics. (source: http://www.infraleuna.de/en/), retrieved 2013-02-11 [translator's note].

9.5 Further Development of the Sulfuric Acid/Water Cryothermal Apparatus

The cryothermal apparatus[89] commenced for the company Siemens and further de-

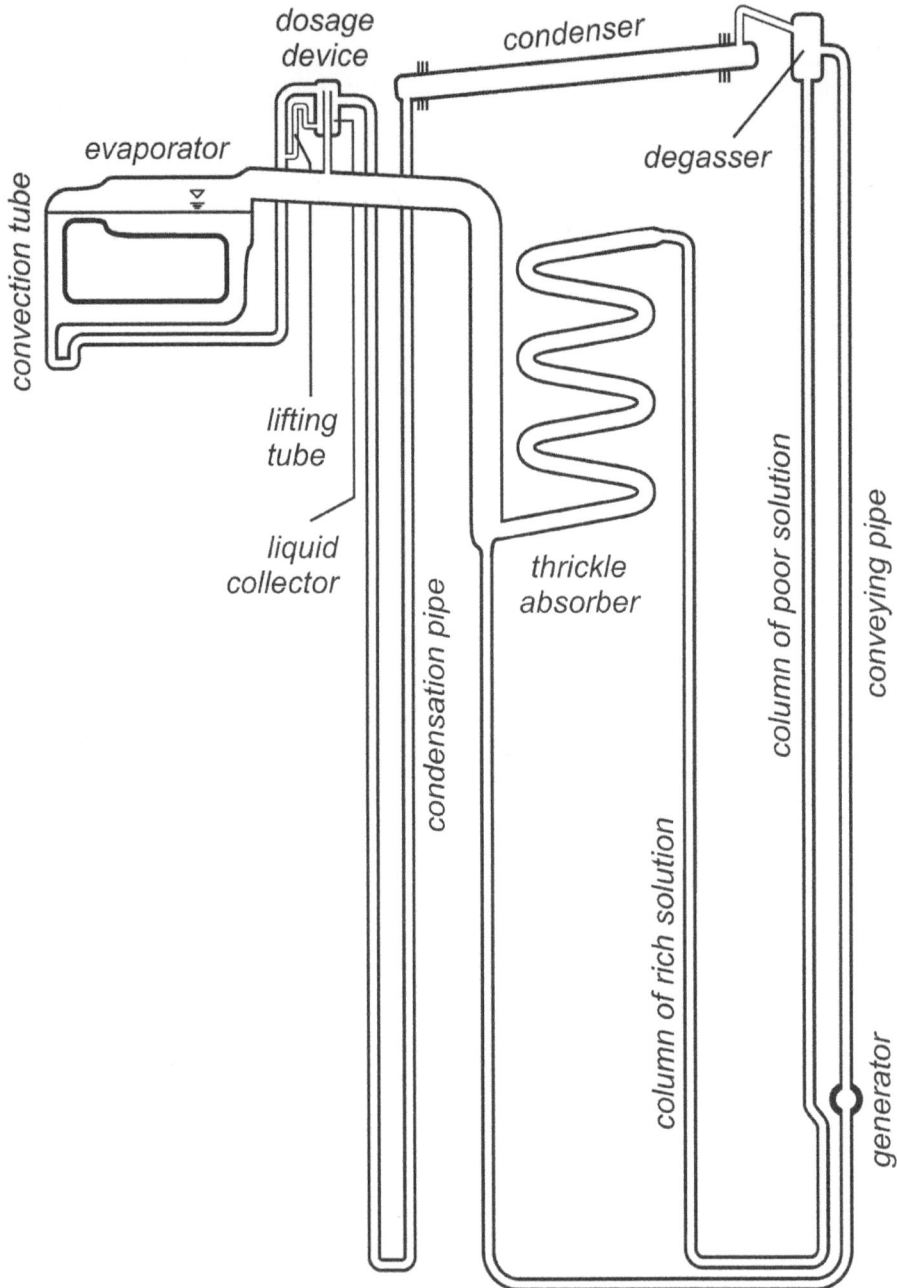

Fig. 9.5: Sulfuric acid/water cryothermal apparatus[164]

---

[89] In favor of the sponsorship of the institute for air and refrigeration technology (ILKA) with the designated manufacturer "DKK Scharfenstein", the commenced development was continued under the leadership of the first author of this report, Siegfried Unger, after Altenkirch's death in Altenkirch's laboratory which, in the meantime, had been attached to the third physical institute of the Humboldt university in Berlin.

veloped by Altenkirch in his laboratory showed technical interruptions due to the carrier gas development after a longer running time.

Due to an appropriate insertion of a droplet pump, loaded with the poor solution, the gas was removed from the end of the absorber[164]. The carrier gas proved to be a mixture of the gases SO_2 and O_2 as the dissociation products of the sulfur trioxide as is described due to the reaction equation $2SO_3 \longleftrightarrow 2SO_2 + O_2$.

The super-heating of droplets at the wall surfaces which were not washed round due to the liquid of the *simmering coil*[164], cf. Figure 6.3 (Section 6 "The Cryothermal Apparatus—A Motionless Heat Transformer"), was recognized as a cause.

Due to the application of constructive measures with which the gas establishment was caused due to the so-called "still" evaporation in the generator, it could be limited as far as possible, and be removed with the suction volume of the droplet pump[90]. The problem was thereby solved. The "still" degasification ensues from the surface of a heated *solution sump* in the generator to which the thermal heat from the vessel walls was transferred, and due to the convection was transported from there to the surface of the solution sump.

Moreover, measures for the prevention of the freezing *of the evaporator during the degasification of the water in the degasser (evaporator) at temperatures below 0 °C* had to be adopted.

That was achieved due to the introduction of the "still" degasification also in the evaporator together with a pulsated loading (cf. Figure 9.5, Unger, S.[164]). To this end, a sulfuric acid concentration of at least 5 % was perpetuated in the evaporator. To this end, the volumes of the solutions V_D in the degasser and V_G in the generator were appropriate to dimension and to install an overflow of the solution from the degasser into the absorber (without blocking the gas way, cf. Figure 9.5): The stabilization of the concentration interval in the degasser due to the return of the solution into the absorber is shown as follows:

- When the generator absorber system is too poor in solvent, the refrigerating capacity drops. The degasser becomes richer in water content and due to the overflow into the absorber the absorber generator system becomes richer in solvent so that the refrigerating capacity increases again.
- When the generator absorber system is too rich in solvent, the refrigerating capacity rises. The degasser becomes lower in water content since less condensate is supplied than evaporated. As a result, the solution in the degasser becomes richer in solvent, the vapor pressure of the water drops, so that the refrigerating capacity again decreases.

A very effective film absorber was achieved for this cryothermal apparatus due to the loading of vertical tubes with spherical thickenings with the solution from above, in which the solution trickles down inside the interior tubes which are connected in parallel from the gas side. The tubes are loaded at the gas side from below, the residual gas is supplied to the droplet pump through capillaries. The alternate nar-

---

[90] The gas which is steadily collecting at the end of the condenser is also supplied via a very narrow capillary to this droplet pump (not outlined as well).

Several droplets pass a capillary tube and effectuate like a piston which pushes the gas ahead; the disturbing residual gases are sucked off by the pump, in this compressed, and the refrigerant vapor condensed out. The residual gases are collected in a vessel.

rowing and widening of the cross section promotes the establishment of an equably azimuthal film thickness of the solution during the downward trickling.

As to the solution cycle, an equable flow rate of the *thermosyphon pump*[53] at alternating pressure differences between condenser and absorber is supported due to the realization of a bypass at the conveyor pipe of the thermosyphon conveyance tract (Unger, S.[163]).

The work at this was finalized by the first author of this report, Siegfried Unger, so far as to achieve a sample of proven functionality in terms of the refrigeration technology.

The putting into practice due to the institution in charge of the project, ILKA , was —after erection of the prototype and proven perfect operation—without explicit opinion—suspended, presumably due to concerns in respect to glass as building material.

The processes take place as follows: The liquefied working medium passes from the condenser via a U-shaped condensate line into a liquid collector which is connected via a conveying tube with the gas space of the evaporator and absorber. In the liquid collector, initially the condensate level rises so long until the small lifter tube overflows and the convection tube of the evaporator due to the drainage is loaded with the liquid contents of the liquid collector in a short period of time.

The condensate is then evaporated from the surface of the solution in the evaporator with a certain refrigerating capacity. The vapor passes into the trickle absorber, where it is absorbed due to the entering poor solution. The solution enriched with the working medium leaves the absorber through the column of the rich solution and passes into the generator; there the working medium is again driven off due to the heat supply. The developing vapor is transported through the conveying tube into the gas separator from which the poor solution streams into the absorber and the working medium vapor flows into the condenser.

10 Appreciations and Positions

Awarding of the Doctorate of Engineering honoris causa

In the year 1930 Edmund Altenkirch was awarded, in appreciation of his outstanding merits in the field of the reversible heat and cold generation, especially in motionless and multistage absorption refrigeration machines, the academic degree of Doctor of Engineering honoris causa by the Technischen Hochschule Karlsruhe, Germany.

Presidency of Commission VII, Chief Editorship of the Periodical "Zeitschrift für die gesamte Kälte-Industrie"

From 1933 until 1945 Altenkirch was editor-in-chief of the periodical "Zeitschrift für die gesamte Kälteindustrie". After 1936 he was a president of the Commission VII for Research and Education of the Institute International du Froid (Paris).

Award of the Linde-Memorial-Medal

In the year 1950 he was awarded as a first bearer of the Linde Memorial Medal, the supreme award of Der Deutsche Kältetechnische Verein[91] (Figure 10.1).

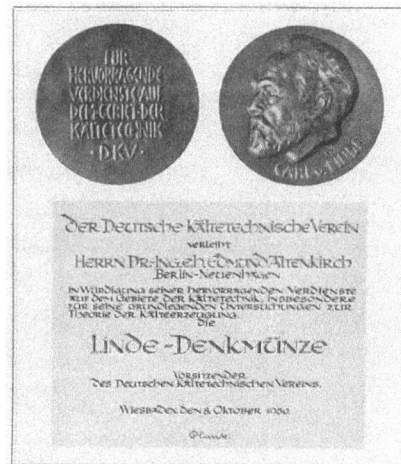

Fig. 10.1: Reading out of the certificate of the bestowal of the Linde-Denkmünze, (from left to right: Dr. Ruppel, Prof. Plank, Prof. Nesselmann, Dr. Altenkirch)

Acknowledgment of Altenkirch's 70th birthday by Prof. Nesselmann

From his former collaborator, Dr. Nesselmann, who participated in the work on the cryothermal apparatuses for the company Siemens-Schuckert-Werke since 1925, the following acknowledgment was publicized (excerpts):

- The German refrigeration engineers claim Altenkirch as one of their first representatives. However, one would be consistent with his unique personality only to a very limited extend to see his feats only in the field of refrigeration technology, although the key area of his professional activity undoubtedly lay and still lies in this field.

---

[91] German Society of Refrigeration and Air Conditioning (DKV) [translator's note].

- Countless are the tasks Altenkirch was occupied with, by far not all have been published. He worked, to name just a few, in the field of the steam boiler engineering, centrifugal pumps, and the mammoth pumps. He contributed to the theory of pumps and compressors, and worked on technochemical problems. He had significant involvement in the development of the steam jet ejector chiller according to Josse and Gensecke, in the determination of an international unit of the refrigerating capacity, in numerical analysis, and in the question of the heat transfer and the pressure drop with turbulent and laminar flow in the gap.

Obituary by Prof. R. Plank

This is published in the foreword of Altenkirch's personally written autobiography published by the German Refrigeration Association, Karlsruhe, shortly after Altenkirch's death and is reprinted below in its entirety:

Edmund Altenkirch ranked far beyond Germany's boundaries as one of the leading experts in the field of refrigeration technology. Since the year 1912, in which he publicized his fundamental works on reversible absorption machines, he enriched in constant progression the scientific refrigeration technology with numerous investigations. He had profound knowledge in physics and outstanding mathematical skills. As a pupil of Max Planck's he trained himself, above all, as an excellent thermodynamic expert. However, he did not content himself with purely theoretical knowledge but showed a vivid interest in thermo-technical problems with which he strived for enhancing the quality grade of energy conversions. His challenge were the irreversibilities, and he always attempted to make the thermal processes as reversible as possible.

As discoverer he was allotted with a tragic fate. Although, due to the close collaboration with influential companies (Siemens-Schuckert-Werke and Borsig), he seemed to have the best perspectives implementing his fruitful ideas into reality, he was frequently stripped of the fruits of his labor due to two world wars and lengthy patent disputes. Since he moreover dedicated himself to the most difficult tasks, he was not understood by his contemporaries. Most of his discoveries will only reach maturity in the future.

The present autobiography, which the German Refrigeration Association dedicates in almost unchanged form to Altenkirch's numerous friends, closes with the year 1949, but Dr. Altenkirch continued his work unremittingly until his death on November 28, 1953 during which time the following works were released: [19] to [79].

Who was lucky to know Edmund Altenkirch personally did not only admire him as a good expert but, above all, they highly appreciated him as a person. His constant cooperativeness and humor, also in difficult situations, his sincerity, and his appreciation for the others' feats were the character of his nature. During leisure hours he wrote poems and other works of literature which he however shyly concealed, and only showed his closest friends. They are imbued with deep earnestness.

The Technical University in Karlsruhe awarded to him the doctor's degree honoris causa in engineering. The German refrigeration association presented to him, as the first, the Linde Memorial Medal.

His former working areas are now used as a research laboratory for the Berlin Humboldt University in which, above all, his unfinished work was continued.

The refrigeration engineers of the whole world owe to him gratitude and will honor his memory.

sgd. Rudolph Plank

Reminiscence on the 10th Anniversary by Prof. Dr. R. Plank

(full wording)

'*Reminiscence to Edmund Altenkirch*

It is ten years ago, since Edmund Altenkirch has past away. We honored his memory when he had been seventy years old on August 11, 1950. Professor K. Nesselmann praised him as a scientist, as an engineer, and as an individual. His great multiplicity was clearly expressed in this appreciation. We also had expressed our sorrow at his death in this journal[101], in the December issue of 1953. Since then, ten years have elapsed, which is a long lifespan in our fast-changing time, in which such a huge volume of news broke about us that we slightly forgot which had gone before. However, the work of Altenkirch must not be forgotten and the same applies to his humane personality. Regarding his manifold fields of science which he processed, the absorption machines are in the foreground. One of his many pupils in this field was the recently deceased Wilhelm Niebergall who had revised the seventh volume of the handbook of refrigeration technology, and gave in this the most comprehensive coverage of the development of the absorption machines; in the index of names of this volume Edmund Altenkirch was cited 36 times.

Although Altenkirch had not lectured at technical universities, he gave direction to many young engineers with which he met at industrial plants and in everyday life, and influenced them permanently. He radiated a large amount of knowledge, one received kind-hearted and tolerant instruction from him, and also the sense of humor was not extraneous to him.

Although he could not achieve major commercial successes, at least his work was highly valued due to the bestowal of a doctor's degree honoris causa, and the golden Linde *Memorial Medal awarded by the German refrigeration association.*

Only a few know that Edmund Altenkirch was also a poet, he always wrote poems in his youth, and his verses became more and more serious as time went by but never unhopeful and negating. Besides shorter poems, he also wrote a play as well as a drama in verses, which he would like to have publicized but he abstained from the search for a publisher after the first failure. In his belongings several things of it were found, which was not really known prior even to his closest relatives[92].

To revive the memory of him continually updated is our gratitude for his pioneering deeds and our duty toward the younger generation.

On his tombstone could be written the words of the Russian poet S. J. Nadson (1886) who passed away at a young age:

> *Don't say he passed away—no, he is alive,*
> *Destroyed is the altar—the flame continues flaring up,*
> *The rose plucked—remains framed by its perfume,*
> *The harp broke in two—the chord resounds still![93]*'

---

[92] His theater play bears the title "The Eremite"; further poetry, inter alia a small volume of love poems, are in the archive of his grandson Wolfgang Altenkirch in manuscript form[46].

[93] Transcribed from the German of the author into English by Michael Unger [translator's note].

11 Appendix

11.1 Contributions to Gehlhoff's Handbook of "Technical Physics"[94]

The contributions to Gehlhoff's handbook[22] are exemplary in focusing on the technical aspects and, especially for the practical use, on the most important. Therefore some contributions shall be expressed here from the chapters 12.1 - 12.7 in excerpt form, together with the introduction to his developments for open systems of air treatment (chapter 6.4).

11.1.1 Principle of the Steam Jet Ejector Chiller

Regarding the steam jet ejector chillers the compression of the vapor is caused by flow processes. Compared to refrigeration machines with a steam compression system they have the benefit of using hydrogen as working fluid. To this end the benefit of the economy of space and the exclusion of moving machine parts are also facts that make this type of machine appropriate for some fields of application despite the disadvantage of a larger steam consumption.

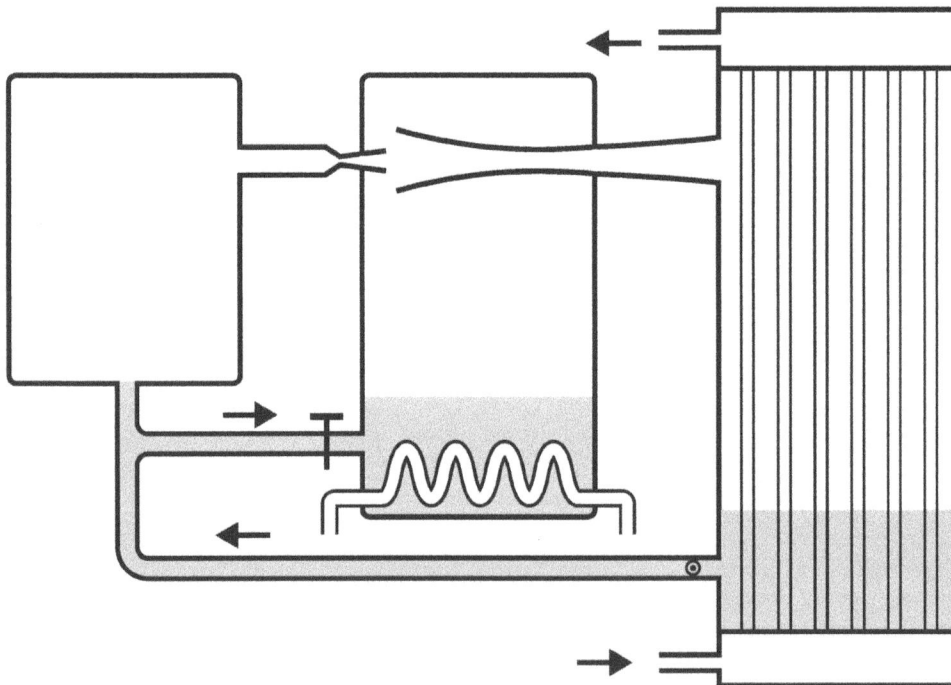

Fig. 11.1: Principle of the steam jet ejector chiller

In Figure 11.1 the vapor source is a steam boiler (outlined here), or an exhaust steam accumulator. The vapor depressurizes in the nozzle, cools down[95], and gathers speed so that it carries vapor from the evaporator away. In this, brine is located

---

[94] Today in English it is often called 'engineering physics,' added by David Stover [translator's note].

[95] Supplementation according to the first author Siegfried Unger

when temperatures below 0° C shall be achieved. The vapor mixture is conveyed thereupon into the diffuser, in this the high speed is converted into pressure again. To this end, the vapor is severely superheated and eventually condensed in the condenser. The condenser is cooled as outlined due to the cooling water.

The cold is absorbed due to the condensing coil in the evaporator and conducted to the place of consumption. Regarding the type Josse-Gensecke the brine is sucked off from the evaporator itself and conducted to the cold storage rooms and enters through a regulating valve (not outlined) again into the evaporator. Regarding the type Leblanc several rims of the nozzles belong to a diffuser.

The brine is continually more concentrated due to the suction of the water vapor from the brine, and this evidence would soon cause salt development and disruption of the operation unless the supplementation of water was provided continuously. In the drawing, therefore the pipeline from which the water is recycled from the condenser into the steam generator is branched off from the loading line (with the condensate) which supplies the missing water through the regulating valve again back to the evaporator.

According to Altenkirch's approximate theory it results for the live steam a heat ratio between 0.3 and 0.4, for the exhaust steam a ratio of about 0.25. The super-heating of the vapor entails an improvement of about 25 %.

11.1.2 Air Refrigerating Machine (Excerpt)

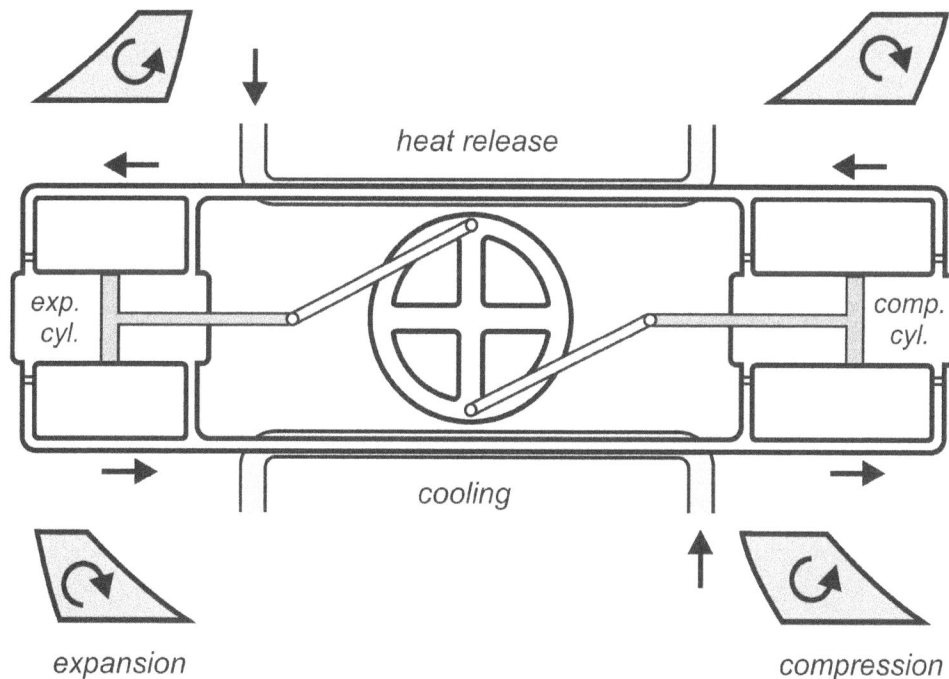

Fig. 11.2: Air Refrigerating Machine

As in Figure 11.2 shown, in the closed air refrigerating machine according to Ericsson the air is decompressed in essence adiabatic in the compression cylinder from the initial pressure p_1 and the initial volume V_1 upon the final pressure p_2 and the volume V_2, whereby the temperature grows from T_1 up to T_2'.

It is then pushed due to the emission of heat to a cooling medium (cooling water) with the constant pressure p_2 into the expansion cylinder. To this end, the temperature drops to T_2 and the volume to V_2.

In this, the air depressurizes, in essence, adiabatic until the pressure p_1 and the volume V_1, whereby the temperature drops to T_1. It cools then a medium (brine) due to a heat withdrawal. After that, the air is again transported into the suction chamber of the compression cylinder. To this end, the temperature rises with it from T_1' to T_1, and the volume from V_1' to V_1.

With the working diagrams which are plotted over the compression cylinder and the expansion cylinder, the process in the machine may be followed comfortably. The work for the refrigeration capacity to be used is the difference between the compression work in the compression cylinder and the expansion work in the expansion cylinder.

Altenkirch considers the efficiencies η_c and η_e of the compression and expansion processes as well as to this end the accompanying decrease of the refrigerating capacity, and comes to the expression

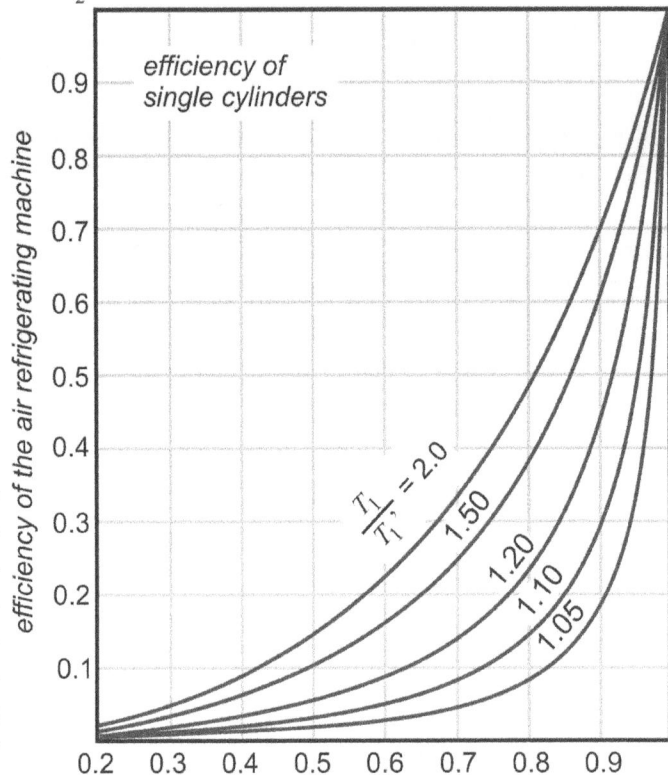

Fig. 11.3: efficiency of the air refrigerating machine

$$\eta' = \frac{\dfrac{T_1'}{T_1} - 1}{\dfrac{T_1'}{T_1} \cdot \left(\eta_c \eta_e\right)^{-1} - 1}$$

(Eq. 11.1)

Let the efficiencies of the working cylinder be η_c and the expansion cylinder η_e the same size, so η' can be specified as the function of η_c, η_e, as shown in Figure 11.3.

Hence, it follows that for small temperature intervals of the warming-up and cooling-down phase of the air, the quality grade η' between the cylinders turns out to be extraordinarily little, even if the single quality grade of the cylinder is higher. For a temperature interval of ca. 30 K, that means for $T_1/T_1' = 1.1$, the quality grade $\eta' = 0.27$ when $\eta_c = \eta_e = 0.9$—this is one of the reasons why the air refrigerating machine is not applicable for the practice. Another reason is the size of the compressor dimensions and heat transfer surfaces. The latter evil can be reduced very easily due to the application of very high pressures of 50...100 kbar ("Gesellschaft für LINDES Eismaschinen")[96].

11.1.3 Air Liquefaction According to Linde (Excerpt)

'The Joule-Thomson Effect is the basis for the air liquefaction according to LINDE and thus to the cryogenic engineering and its main application field the gas separation. In the Figure, the schematic of a machine with two cycles is outlined: a high-pressure cycle and a low-pressure cycle. Due to the low-pressure compressor, fresh air from the atmosphere is aspirated after a preceded drying through a connecting piece and the transportation into the high pressure compressor.

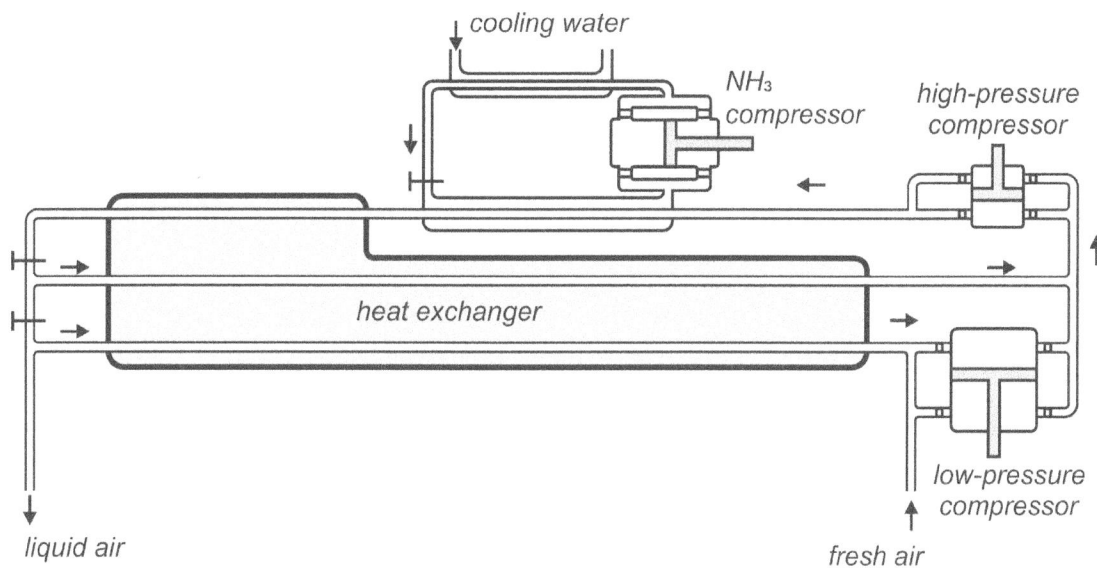

Fig. 11.4: High-pressure cycle and precooling in the Hampson–Linde cycle

In this, the air is compressed to a still higher pressure and is then transported through a heat exchanger, in this precooled, and then to a throttle point.

Here, it decompresses and is thus partly conveyed through a heat exchanger again back to the high-pressure compressor, from where it is again condensed together with the arriving air from the low-pressure compressor. Another part is decompressed in a second throttle point, and the established liquid air can be extracted. Partly, the air depressurized upon the atmospheric pressure is again transported back through the heat exchanger into the low-pressure compressor where it is aspirated and decompressed together with the fresh air of the atmosphere.

Both compressors are kept as cool as possible by the cooling water. Since vapor-compression refrigeration systems[97] are working considerably more favorable than cold air machines so larger facilities normally apply a precooling of the air which is compressed upon a highest pressure in the high-pressure cycle, viz to the lowest

---

[96] Founded by Carl von Linde in 1879. Retrieved from wikipedia.org 2013-03-28 [translator's note].

[97] Jacob Perkins (1766-1849) (source: Modern Engineering Thermodynamics by Robert T. Balmer), retrieved 2013-03-26 form google books, p. 543 [translator's note].

temperatures which are still comfortably achieved with compression machines. This precooling is indicated in the Figure [11.4] as well.'

11.1.4 Air Liquefaction According to CLAUDE

'The method according to Claude is distinguished from the method according to Linde only in that there is a working cylinder instead of the throttling in the high-pressure cycle of which lubrication encountered difficulties in the past; yet there are now oils known, which come here into question, which maintain their lubricity at low temperatures. Theoretically, the method according to Claude is superior to the method according to Linde since in a working cylinder the reversible adiabatic relaxation takes the place of the irreversible throttling. Practically, there is however hardly a difference in the need of work for the generation of the liquid air in the different working procedures since the quality grade of the expansion cylinder impairs severely the theoretical perfectibility of the process whereas the Hampson-Linde cycle has the advantage of a greater simplicity.'

11.1.5 Gas Separation and Rectification

'The main field of application of the liquid air is, besides its use

Fig. 11.5: Rectifying column

for blowing-up in mines, the gas separation, that means the extraction of pure nitrogen and oxygen from the air whereby also the rare gases argon, neon, and the like, can be produced.

The Figure 11.5 shows a rectifying column in which the complete fragmentation of the liquid air into oxygen and nitrogen is executed, as well as a two-column apparatus with a lower rectifying column and an upper rectifying column. The compressed air enters, as outlined in the accompanying sketch, the counter-current apparatus.

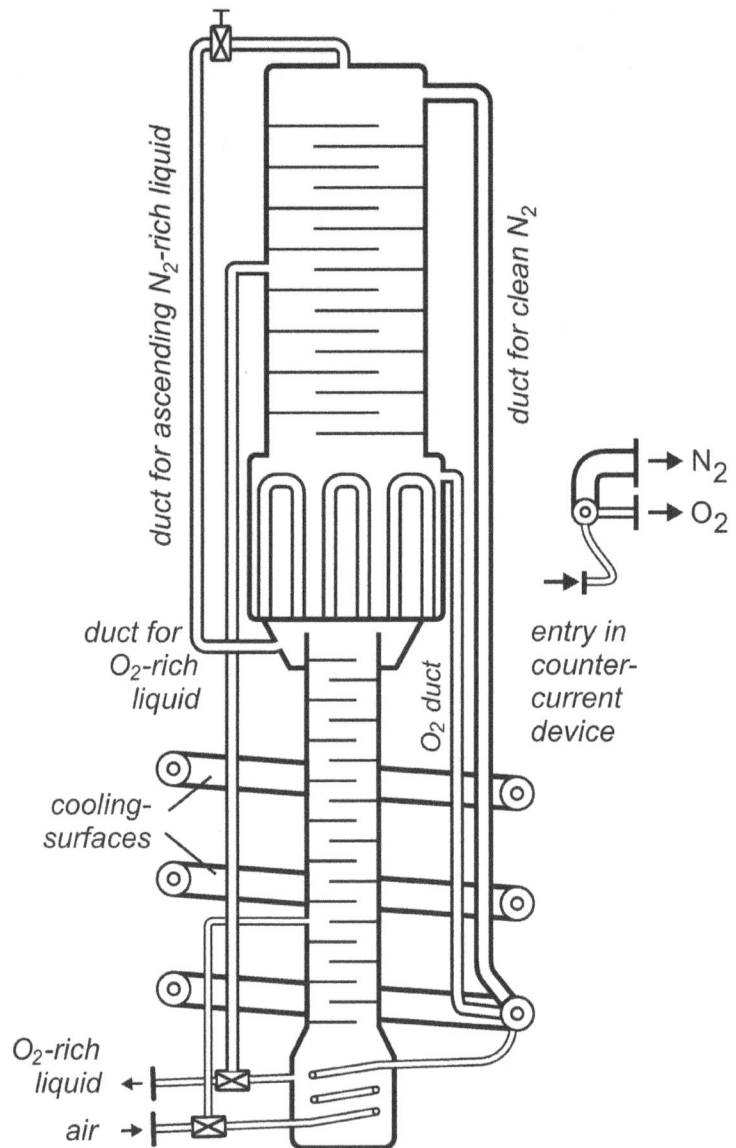

This consists of two pipes which are stuck into each other. In the one the pure oxygen, and in the annular space of the two pipes the pure nitrogen streams from the rectifying column. The air goes into a narrow, spiral shaped pipe throughout the annular space until the lower throttle valve. It streams then into the ascending pipe and enters at the level of the lower third into the lower rectifying pipe. Here, it trickles down, and is vaporized at the bottom due to heat contact with the sump of the precooled compressed air which serves here as the heating liquid. The vapor becomes more enriched in nitrogen during ascending due to the condensation of oxygen at the liquid air which trickles down. Since the lower rectifying column has an overpressure of 5 - 6 kbar, the residual nitrogen condenses at the assembled and here outlined cooling surfaces at the level of the junction of the lower to the upper rectifying column.

There, as outlined, the nitrogen-rich liquid is collected, and is introduced, after passing the upper reducing valve from the horizontally outgoing line, as trickling liquid into the upper rectifying column.

The liquid air, which has collected in the lower rectifying column at the bottom due to the trickling-down, and has become enriched in oxygen due to the evaporation of nitrogen and the uptake of oxygen, is introduced from the surface of this bottom sump via the second regulating valve, assembled at the height of this surface, and its line ascending from this, as trickling liquid into the middle of the rectifying column.

The liquid, which is trickling-down in the upper rectifying column, is evaporated at the upper side of the cooling surfaces due to the condensation heat of nitrogen which is transferred there from the lower rectifying column. It consists of pure oxygen since it has already delivered nitrogen at an increasing rate, and is introduced through the shortly above the cooling surfaces to the right-hand side and downwardly outgoing pipe into the counter-current heat exchanger.

Concomitantly, the gas mixture, which is ascending in the upper rectifying column, is enriched in nitrogen due to the nitrogen-rich liquid which is trickling down, which goes through the pipe which is going out to the right-hand side and downwardly into the counter-current heat exchanger which is thus be loaded with pure nitrogen, and like the pure oxygen at the nozzle (O_2), and can be taken from the nozzle (N_2), outlined accompanying to the sketch.

The mutual washing-out of the nitrogen and oxygen occurs in the rectifying column at a great speed, and takes only a few seconds. Also, the re-heating of the nitrogen and oxygen from the boiling point until the room temperature does only take time of approximately 1 second so that the whole process proceeds extraordinarily fast, and the rectifying columns can also be designed comparably small for the production of large quantities of nitrogen and oxygen.'[98]

---

[98] Descriptions and graphic displays are supplementarily processed by the authors due to the lack of originals.

11.2 Contemporary Witness of the Work after World War II
Two letters by Altenkirch to his wife, 1946

The following ride to Zwickau, the place of his later work for the SMAD[99], Altenkirch, whose left arm was missing due to a tragic accident (to this end, also see Section 2), had undertaken solely.

First letter,

'Dear Margarete,   Dresden, February 11,1946, 11:05 P.M.

I sit here in the overcrowded waiting room of the Dresden-Neustadt station[100] on my trunk next to the central heating. There is a heavy draft since through the adjacent door the passengers are permanently entering and leaving.

On the move, I had interesting talks with the Siemens engineers, from which I also found out some essentials about the continuation of the journey. Hotel Demnitz is hopelessly far away, and during the darkness it is virtually not reachable with the local traffic facilities and due to the heavy baggage. Still farther away is the intermediary agent to whom I was recommended by Mister Köhler[101].

So, the only way open to me is to wait throughout the night here from where the train to Zwickau shall continue at 7:20 A.M.

The counter for the required admission ticked is already closed. Moreover, the train is sold out. Hence, I must continue with the regional train which departs from the Dresden main station tomorrow morning at 10 o'clock. However, there is neither a tram, nor a regional train from here to there ...

In the night walking through the ruins on foot, as well as without knowing the way, will be too dangerous due to the great uncertainty. Hence I remain here, also due to the heavy baggage. Hence, the travel proceeds as uncomfortably as it began and like I have expected.

I sit here next to the heating still quite tolerable, the movement which causes the draft slowly abates. And I look at the crowds of home coming solders passing by, and hope that Joachim's (his youngest son, the authors) face might emerge among them ... And I would not know whether I took him along with me to Zwickau or whether I returned with him to Neuenhagen, but I did not part from him ...'

Second letter,

'Zwickau, February 14, 1946, Franz-Mehring-Straße 3-7, at Weissflog's

'Dear Margarete,

Continuing my letter from February 11, 1946, when I was at the waiting room of the Dresden-Neustadt[2] station, I can tell you, that a young wood turner (18 years old)

---

[99] Sowjetische Militäradministration in Deutschland (Soviet Military Administration in Germany [translator's note]).

[100] second largest railroad station in Dresden, [translator's note], retrieved from wikipdia.org, 2013-03-28

[101] The Owner of the company Kälte-Köhler in Zwickau [translator's note].

from Görlitz[102] carried my baggage to the tram stop at 5:00 A.M. There, I had to wait rather long during the increasing cold and a bit of snow blowing, until it arrived. Unfortunately, there was still a walk of a quarter of an hour from its terminal necessary to reach the main station.

Hence, I dragged my trunk step by step through the ruins among a stream of people who also hastened to the main station. Although I was there abundantly early, it was merely impossible to even conquer a standing place at the long train, and I had to stay back with about 100 other travelers.

Hence, I sat in the waiting room to wait for the next regional train by 2:42 P.M. and I ate a plate of beet soup ... Three hours before the departure of the train I went to the platform which was already full of people. It was very cold and windy with strong snow drifts. So I got on a train on the side track and waited until my train came in ...

Around 12:30 P.M. the train arrived, and the most terrible rush started to get into it. I was lucky and gained a seat by putting my trunk next to the toilet door and sat next to it.

The compartment got more and more crowed. Eventually, I could not stir a foot and had to duck the head, initially in this fettle I had to persevere two hours until the departure of the train and then five additional hours until Chemnitz[103].

In Chemnitz it got cleared, and one could use the lavatory again, into which five people with a lot of baggage had climbed through the window. To this end, my female neighbor badly squeezed two fingers of a female passenger ...'

Fig. 11.6 shows the first lines of the second letter to his wife.

Fig. 11.6: Autograph of the beginning of the letter

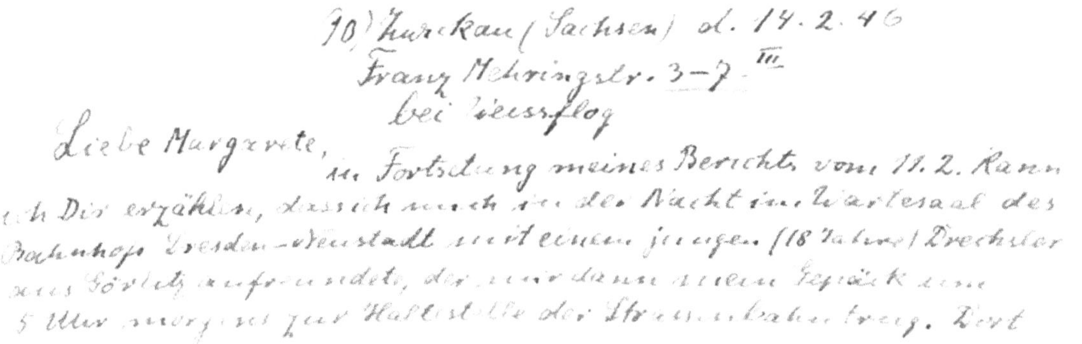
10) Zwickau (Sachsen) d. 14.2.46
Franz Mehringstr. 3-7. III
bei Weissflog
Liebe Margarete, in Fortsetzung meines Berichts vom 11.2. kann ich Dir erzählen, dass ich mich noch in der Nacht im Wartesaal des Bahnhofs Dresden-Neustadt mit einem jungen (18 Jahre) Drechsler aus Görlitz anfreundete, der mir dann mein Gepäck um 5 Uhr morgens zur Haltestelle der Strassenbahn trug. Dort

Continuation of the letter after the arrival in Zwickau

'After this terrible travel—the train was unlit—we arrived around 9:45 P.M. in Zwickau. A night's lodging in the hotel was not available, all was occupied ... Again I dragged myself—my heavy trunk I had deposited at the baggage check-in—throughout the night and arrived after half an hour at the "Volkshaus" ...

[102] A Saxony town, near to the (Polish) eastern boundary of post-war Germany [translators note].

[103] A town in the south of Saxony [translator's note]

... Here, the porters proved to be cooperative and friendly which I enjoyed twice as much after this unspeakable travel.

... One of the porters found telephonically that a private accommodation was available to which he even guided me so that I did not get lost in the foreign town ...

Early in the morning, on February 13, I first went to the private residence of Köhler[104], before the house stood a Russian guard with a rifle since Mister Köhler's house was occupied by the Russians, and he had been in the meantime forced by them to move out ...

At Kornmarkt 8 my office is situated whereto I asked my way while hobbling ... After the inspection of my mandated office rooms we had lunch in Köhler's private residence ... in the meantime two or three female typists had to leave the room for me, but there the roof leaked and it was still crammed full of materials, and first had to be cleared and cleaned.

... Then, I will make room in my office, big enough for a Russian guard each on the right- and left-hand side next to me can find a place for my constant guarding ...

Then, the typist, Miss Hagedorn, accompanied me to my apartment, Franz-Mehring-Str 3-7, a former health insurance building where I move into a spacious room 2-3 stairways high, looking to the South-East. All is very neat, clean, and friendly. I shall pay solely a rent 85,-- marks for it. The landlady, Mistress Weissflog, is also very helpful and friendly; I am very well cared for ...

Ration cards [for food] I can only receive after an examination by the public health office.

That means, today at 2:00 P.M. I first have to go to the public health office donating a four gram blood sample for typhus examination ... Every louse I do not report, incurs one year's imprisonment ... x-ray examination because of exclusion of tuberculosis ... in the afternoon going to the dermatologist, Dr. Fröhlich. He also had studied in Berlin, we partly had the same teachers (physics, Rubens) ... He knew Leibsch and the Spreewald[105] (my father's birthplace).

Before lunch there was a session with a Russian major who could speak German very well who was for the first time at the "Kältetechnik"[106], and was glad to meet me personally, about whom he had already repeatedly heard in Russia ...

Back to my office, I found a pile of files on my desk which I should work through, with what I started right away. Miss Hagedorn already expected me and handed over the ration coupons, grade 5, to me.

...

Yours, Edmund'

---

[104] The Owner of the company Kälte-Köhler in Zwickau [translator's note].

[105] A part of north-eastern Germany named for "forest at the river Spree," known for its special geomorphology (postglacial moraine landscape, sandy soil), biosphere (lowland, bog habitat), culture (most of the Sorbs, a Slavic minority, live there), and recreation (navigable water canal system for punts) [translator's note].

[106] Company Kälte-Köhler [translator's note].

12 Nomenclature

12.1 Quality Figures and Indexes[107]

12.1.1 Conversions of Heat | Sorption

c	specific heat, cf. Eq. 5.20, [kcal kg^{-1} K^{-1}]
f	a specific solution cycle
$\tau = T_h / T_o$	temperature quotient, cf. Eq. 3.66, 3.67
p	pressure, [bar]
P	wattage [electric power], [J s^{-1}]
Q	heat, in general, cf. Section 5.4 et sequ.,[J = kg·m^2·s^{-2}]
$\dot{Q}$	heat flow, in general, [J s^{-1}]
$\dot{Q}_o$	refrigerating capacity, [J s^{-1}]
$\dot{Q}_h$	thermal or heat output, [J s^{-1}]
$\dot{Q}_{Peltiér}$	Peltiér heat flow, cf. Eqs. 3.2 and 3.4, [J s^{-1}]
$\dot{Q}_\lambda$	heat flow by heat conduction, cf. Eq. 3.11, [J s^{-1}]
$\dot{Q}_\sigma$	Joule heat production, cf. Eq. 3.12, [J s^{-1}]
q	specific thermal output (molar evaporation heat), cf. Eq. 5.2 and Tab. 5.2, [kcal kg^{-1}]
r	isobaric enthalpy of evaporation, definition Eq. 5.5, [kJ·kg^{-1}]
ξ	gas constant, cf. Eq. 5.2
S	entropy, cf. Eq. 5.14, [J K^{-1}]
t	Celsius temperature (centigrade), cf. Fig. 5.2 [°C]
T	absolute temperature, [K]
$\Theta = T^{-1}$	reciprocal absolute temperature, [K^{-1}]
T_m	thermodynamic mean temperature, in general
T_{Gm}	thermodynamic mean temperature of the vapor generation in the generator
T_{Vm}	thermodynamic mean temperature of the evaporation
$T_{(A+C)m}$	common thermodynamic mean of the temperatures of the heat dissipation Q_C of the condenser and Q_A of the absorber
W	work, cf. Eq. 3.5, [J]
ξ	mass fraction, cf. Eq. 5.5
ξ	concentration, cf. Eq. 5.19 and Fig. 6.6 [-]
x_i	initial concentration in the generator, cf. Section 5.7.1
$T_{Gm} - T_{(A+C)m}$	infrequently denoted as "temperature stroke of the system"

---

[107] Based on 2009 ASHRAE Handbook—Fundamentals (SI), Chapter 37, Abbreviations and Symbols (Y10.4-82)[178] [translator's note].

$T_{(A+C)m} - T_{Vm}$ infrequently denoted as "temperature lift of the system"

$COP_0 = \varepsilon_0$ coefficient of performance (COP) of cooling, definition Eqs. 4.1, 4.2), [-]

$COP_h = \varepsilon_h$ coefficient of performance (COP) of heating, definition Eq. 4.2), [-]

η quality grade, in general., definition Eq. 3.47, [-]

η_c efficiency (Altenkirch), working cylinder, cf. Eq. 11.1, [-]

η_e efficiency (Altenkirch), expansion cylinder, cf. Eq. 11.1, [-]

η_{Carnot} Carnot quality grade, definition Eq. 5.4, [-]

η_0 quality grade of refrigeration, definition Eq. 3.46, [-]

η_h quality grade of heat generation, definition Eq. 3.50, [-]

v loss factor = reciprocal quality grade, [-]

$\zeta = Q_0/Q_h$ heat ratio of the absorption machine (ζ = quotient of the utilizable cold Q_0 to the supplied thermal heat Q_h)), [-]

λ thermal conductivity, [W/(m·K)]

12.1.2 Thermoelectric Cold Production

A, B geometry parameters of the thermocouple legs a, b, quotient of the cross-section and length (Altenkirch), [m]

a, b thermocouple legs

$\alpha = A/B$ geometry factor of the thermocouple, [-]

$\mathcal{E}_x$ absolute thermoelectric power of the leg's material X [V K^{-1}]

L_σ electric conductance of the thermocouple legs, [A V^{-1}]

I electric current, [A]

$R = L_\sigma^{-1}$ electric resistance of the thermocouple legs, [Ω]

$L_{\sigma\lambda}$ thermal conductance of the thermocouple legs, [J K^{-1}]

$\mathcal{L}$ constant of the Wiedemann-Franz-Lorenz law, valid for ideal metals: $\mathcal{L} = \frac{1}{4} \cdot 10^{-7}$ V^2 K^{-1}.

$\mathcal{L}_X$ Lorenz variable for whichever material X, [V^2 K^{-1}]

$\mathcal{E}^{-1}$ expense figure, (max. definition Eq.3.81, min. definition Eq. 3.82), [-]

ζ expense figure, cf. Section 3.6, [-]

ρ_x deviation of a material X from the WFL law, limit value for an ideal metal = 1, otherwise $\neq 1$. [-]

σ_{id} electric conductivity for ideal metals

T_h, T_0 temperatures at the hot or cold solder junction[108], [K]

---

[108] [thermojunction][175] [translator's note].

T_{m} arithmetic mean temperature, [K]

σ specific electric conductivity, [$\Omega^{-1}\mathrm{m}^{-1}$]

Π Peltiér coefficient

ϵ resulting thermoelectric power between the legs materials α against b [Seebeck coefficient], [$\mathrm{V\,K}^{-1}$]

ϵ' effective thermoelectric power acc. to Altenkirch, [VK^{-1}]

Ψ efficiency (Altenkirch), [-]

Z figure of merit (Lotz[56]), [$\mathrm{K\,\Omega}^{-1}\mathrm{J}^{-1}$]

v degree of loss of the thermoelectric refrigeration, [-]

v_{o} degree of loss of the thermoelectric energy generation of the thermopile, [-]

12.2   General Terms

a,b designations for cycles, see Eqs. 5.32 through 5.35

A Absorber

A,B, C, D, E referring to quadrangles, see Section 4.1.1

abs referring to absolute

BR referring to brine

$comb$ referring to combined

CW referring to cooling water, cooling medium

c referring to cooling, cold, cf. Fig. 9.2 and 7.1

$comp.$ compression, cf. Fig 11.2

$C, cond$ condenser (liquefier)

$cyl.$ cylinder, cf. Fig. 11.2

D degasser (evaporator of the resorption machine), e.g. cf. Section 5.9 et sequ.

E referring to savings, cf. Fig. 4.7

e referring to final, environmental, exterior

$exp.$ expanding, cf. Fig. 11.2

$gen.$ referring to generator, cf. Fig. 5.23

G generator (expeller, boiler)

h referring to hot, also cf. Fig. 7.1

h referring to heating, also cf. Fig. 9.2

H referring to heating medium, cf. Section 5.7.2 et sequ.

HP Referring to high-pressure. cf. Fig. 5.19

i initial

i stages of the cascade, cf. Eq. 3.66

$inlet\ (in)$ referring to entering

LP	referring to low-pressure, cf. Fig. 5.19
m	referring to thermodynamic mean
m	referring to moisturization, cf. Fig. 9.2
mod	modulo, cf. Section 9.2.2 et sequ.
over	referring to overlapping
outlet (out)	referring to leaving
p	referring to poor solution, cf. Eq. 5.19
r	referring to rich solution, cf. Eq. 5.19
R [r]	referring to center, cf. Section 9.2.2 et sequ.
R	resorber (condenser of the resorption machine)
S	referring to segment, cf. Fig. 4.5
V	evaporator (vaporizer)
V	valve
whole	referring to total
x	weight ratio

12.3 Abbreviations

acc.	according
comp(r).	compression, cf. Fig. 11.2 and 4.4
COP	Coefficient of Performance
cyl.	cylinder, cf. Fig. 11.2
DKV	Deutscher Kälte- und Klimatechnischer Verein
DKK	Deutsche Kühl- und Kraftmaschinen GmbH[109]
DDR	Deutsche Demokratische Republik
D.R.P.	Deutsches Reichspatent
Eq(s).	equation(s)
exp.	expansion, cf. Fig. 11.2
GDR	German Democratic Republic
GfKK	Gesellschaft für Kältetechnik-Klimatechnik mbH, Cologne
ILKA	Institut für Luft- und Kälteanlagen
SMAD	Sowjetische Militäradministration in Deutschland
SSW	Siemens-Schuckert-Werke
VEB	Volkseigener Betrieb
Z. Ges. Kälte-Ind.	Zeitschrift für die gesamte Kälte-Industrie

---

[109] [http://de.wikipedia.org/wiki/Foron] retrieved 2015-04-11 [translator's note].

13 Bibliography

13.1 Papers

1   Altenkirch, E., Über den Nutzeffekt der Thermosäule, Physikalische Zeitschrift, Volume 10, pp. 560 to 568. 1909

2   Altenkirch, E., Elektrothermische Kälteerzeugung und reversible elektrische Heizung, Zeitschrift für die gesamte Kälteindustrie, Volume 19, p. 1. 1912

3   Altenkirch, E., Vortrag, gehalten auf dem Int. Kälte-Kongress in Chicago, 1913

4   Altenkirch, E., Umkehrbare Heizung, Feuerungstechnik, year's issue 1, Number 9, p. 160. 1913

5   Altenkirch, E., Reversible Absorptionsmaschinen, Zeitschrift für die gesamte Kälteindustrie, Volume 20 / 21, pp. 1, 114, 150 / pp. 7, 21. 1914

6   Altenkirch, E., Graphische Ermittlung von Heiz- und Kühlflächen bei ungleichmäßiger Wärmeaufnahmefähigkeit der Wärmeträger, Zeitschrift für die gesamte Kälteindustrie, Volume 21, p. 189. 1914

7   Altenkirch, E., Die Erhöhung der Wirtschaftlichkeit von Heizungsanlagen durch den Einbau von Kältemaschinen, Zeitschrift für die gesamte Kälteindustrie, Volume 25, pp. 49 - 57. 1918

8   Altenkirch, E., Eigenschaften der Chlormagnesiumlösung, Zeitschrift für die gesamte Kälteindustrie, Volume 25, p. 87. 1918

9   Altenkirch, E., Eigenschaften der Chlornatriumlösung, Zeitschrift für die gesamte Kälteindustrie, Volume 26, p. 49. 1919

10   Altenkirch, E., Eigenschaften der Chlorkalziumlösung, Zeitschrift für die gesamte Kälteindustrie, Volume 26, p. 77. 1919

11   Altenkirch, E., Das Nachfüllen von Salz, Zeitschrift für die gesamte Kälteindustrie, Volume 27, p. 8. 1920

12   Altenkirch, E., Reversible Wärmeerzeugung, Zeitschrift für technische Physik / Berg- und Hüttenmännischer Tag 117, Volume 1, p. 77 and p. 93. 1920

13   Altenkirch, E., Reversible Wärmeerzeugung, Zeitschrift für technische Physik, Number 4 and 5. 1920

14   Altenkirch, E., Erzielung und Nutzbarmachung hoher Kühlwasserablauftemperaturen bei Kompressionskältemaschinen, Zeitschrift für die gesamte Kälteindustrie, Volume 28, p. 93. 1921

15   Altenkirch, E., Neue Dampftabellen für Ammoniak, Zeitschrift für die gesamte Kälteindustrie, Volume 23, p. 173. 1921

16   Altenkirch, E., Grundlagen und Methoden für die Berechnung von Leistungstabellen für die Kompressionskältemaschinen, Zeitschrift für die gesamte Kälteindustrie, Volume 29, p. 163. 1922

17   Altenkirch, E., Kälte als Energiespeicher, Volume 30. 1923

18   Altenkirch, E., Die neue Luftkältemaschine nach Maurice Leblanc, Zeitschrift für die gesamte Kälteindustrie. 1923

19   Altenkirch, E., Die Nutzbarmachung der Überhitzungswärme bei Kompressionskältemaschinen (Vortrag), Zeitschrift für die gesamte Kälteindustrie, p. 57. 1923

20   Altenkirch, E., Beiträge zur Theorie von Pumpen und Kompressoren, Zeitschrift für technische Physik, Volume 5, Number 2. 1924

21   Altenkirch, E., Beitrag zur Theorie von Pumpen und Kompressoren, Zeitschrift für technische Physik, Volume 5, pp. 44 to 53. 1924

22   Altenkirch, E., Kältetechnik, Beitrag zu Gehlhoffs Lehrbuch der Technischen Physik, Volume 1, pp. 362 - 365, Johann Ambrosius Barth, Leipzig, 1929

23   Altenkirch, E., Report on the Work of the Special International Committee On The Practical Refrigeration Unit And Test Code, Annexes to the Bulletin of the International Institute of Refrigeration, 11th. Series, Number 4. 1936

24   Altenkirch, E., Rudolf Plank zum 50. Geburtstag als Hrsg., Zeitschrift für die gesamte Kälteindustrie, Volume 43, Number 3, p. 47. 1936

25   Altenkirch, E., Bericht über die Arbeiten der Kommission für Kälteeinheit und Leistungsregeln, VII. Internationaler Kältekongress. 1936

26   Altenkirch, E., Lufttrocknung durch Kühlung, Wärme- und Kältetechnik 39, Number 6, p. 2. 1937

27   Altenkirch, E., Neue thermodynamische Wege zur Luftbehandlung, Zeitschrift für die gesamte Kälteindustrie, Volume 44, p. 110. 1937

28   Altenkirch, E., Der Anteil der Firma Borsig an der Entwicklung der Kältetechnik, Zeitschrift für die gesamte Kälteindustrie, Volume 44, pp. 2, 13. 1937

29   Altenkirch, E., Neue thermodynamische Wege zur Luftbehandlung, Zeitschrift für die gesamte Kälteindustrie, Volume 44, p. 110. 1937

30   Altenkirch, E., Die Trocknung schwach hygroskopischer Stoffe, Wärme- und Kältetechnik, 40, Number 6, p. 81. 1938

31   Altenkirch, E., Plank, R., Über die Möglichkeit der Festlegung eines internationalen Vergleichsprozesses von Kältemaschinen und einer internationalen Einheit der Kälteleistung, Zeitschrift für die gesamte Kälteindustrie, Volume 45, p. 121. 1938

32   Altenkirch, E., Trocknungsanlage mit Ausnutzung der Sonnenenergie, Z. VDI,

Volume 82, pp. 1347 and 1348, Z. VDI, Volume 82, pp. 1347, 1348. 1938

33 Altenkirch, E., Teilwertrechnung, Zeitschrift für die gesamte Kälteindustrie, Volume 46, p. 1. 1939

34 Altenkirch, E., Wärmeübergang und Druckverlust bei turbulenter und laminarer Strömung im Spalt, Zeitschrift für die gesamte Kälteindustrie, Volume 51, p. 1. 1944

35 Altenkirch, E., Der Aktionsradius ortsbeweglicher Kältespeicher, Die Kälte, Number 1, p. 3. 1948

36 Altenkirch, E., Der Aktionsradius ortsbeweglicher Kältespeicher, Die Kälte, Number 1, p. 3. yr. 1948

37 Altenkirch, E., Die Kompressionskältemaschine mit Lösungskreislauf, Kältetechnik, Volume 2, pp. 251 - 259, 279 - 284, 310 - 315. 1950

38 Altenkirch, E., Der Einfluss endlicher Temperaturdifferenzen auf die Betriebskosten von Kompressionskälteanlagen mit und ohne Lösungskreislauf, Kältetechnik, Volume 3, pp. 201 - 205, 229 - 244, 255 - 259. 1951

39 Altenkirch, E., Über die technische Bedeutung des inversen Thomson-Joule-Effektes im kritischen Bereich, Allgemeine Wärmetechnik, Volume 2, pp. 121 to 123. 1951

40 Altenkirch, E., Schnellläufige Regeneratoren, Verlag Technik, Berlin. 1952

41 Altenkirch, E., Verzögerungsfunktion, Verlag Technik, Berlin. 1952

42 Altenkirch, E., Eine allgemeine Gleichung für den Wärmeübergang im glatten Rohr, Kältetechnik, Volume 5, pp. 253 and 254. 1953

43 Altenkirch, E., Klimaregelung in Kühlräumen, VEB Verlag Technik, Berlin. 1954

44 Altenkirch, E., Absorptionskältemaschinen, VEB Verlag Technik, Berlin. 1954

45 Altenkirch, E., Mein Lebenslauf 1880-1953 (with a foreword by Rudolph Plank), C.F. Müller, Karlsruhe. 1955

46 Altenkirch, E., Writings and correspondence of Altenkirch [in German], assets. Commentary[110]: The assets are accessible at Wolfgang Altenkirch's, graduate mathematician, and Dr. Christel Altenkirch, attorney-at-law, in Neuenhagen near Berlin, Bahnhofstr. 15. 2003

47 Alefeld, G., Wärmeumwandlungssysteme, TU München. 1983

48 Alefeld, G., What needs to be known about fluid pairs to determine heat ratios of absorber heat pumps and heat transformers, Proceedings of the 1987 International Energy Agency Heat Pump Conference, Orlando Florida, Chapter 26, April 28-30, p. 375. 1987

---

[110] In this submitted report all commentaries are written by the first author, Siegfried Unger

49   Alefeld, G., Probleme mit der Exergie, Brennstoff-Wärme-Kraft, Volume 40, pp. 72-80. 1988

50   Alefeld, G., Second Law Analysis for Absorption Heat Pumps and Heat Transformers, Proceedings of a workshop held in London 12 to 14 April, organized by the European Communities and British Gas plc Watson House Research Station. 1988

51   Alefeld, G., The Coefficient of Performance (COP) of Thermal Power Stations Derived from the Second Law, J. Non-Equilibrium. Thermodynamics 16, pp. 153 – 173. 1991

52   Alette, M., Mulder, Novel adsorption heat pump based on metal-supported zeolites for more efficient heating and refrigerating, Project Reference, JOU20437, EU Research Project "Joule". 1996

53   Behringer, H., Die Flüssigkeitsförderung nach dem Prinzip der Mammutpumpe, Dissertation, Technische Hochschule Karlsruhe 1930. 1930

54   Beretteneff, A., A New development in Absorption refrigeration, Refrig. Engineering, Volume 57, p. 553. 1949

55   Yau, M. K. and R.R. Rogers, Short Course in Cloud Physics, Editor: Butterworth-Heinemann. 1989

56   Cube, L. von, Lehrbuch der Kältetechnik, 4th Edition. Heidelberg, Contribution by Helmut Lotz, p. 189, C. F. Müller. 1997

57   Dannies, H., Über Diffusionskälteanlagen, Archiv für die gesamte Wärmetechnik, Volume 1, p. 143. 1950

58   MacDonald, D. K. C., Thermoelectricity, p. 8, equation (2) et sequ., Wiley 1962. 1962

59   Erdmann, Erika, née Altenkirch (daughter), Humankind Advancing, A Quarterly, ISSN 1372335, R.R.1, Lockport, N.S., register number: 08798 Canada

60   Gerlich, A., Gekoppelter Wärme- und Stofftransport in der kompakten Zeolithschicht einer Adsorptionswärmepumpe, Verlag Shaker, Aachen. 1993

61   Goldsmid, H. J. and R.W. Douglas, Elektrothermischer Kälteapparat (Hersteller General Electric Co. Ltd., Wembley), Brit. J. Appl. Physics, Volume 5, p. 386 (mind the later correction of the authors!). 1954

62   Grabenheinrich, H.B., Schulte, U., Die Diffusions-Absorptions-Wärmepumpe auf dem Weg in den Markt, Gas, January/February 2001, pp. 41 - 44. 2001

63   Grassmann, P., Edmund Altenkirch 70 Jahre alt, Kältetechnik, Volume 2, Number 8, p. 181. 1950

64   Groll, E.A., Experimentelle und theoretische Untersuchungen von Kompressionsmaschinen mit Lösungskreislauf, DKV Status Report No. 44, DKV, R23-DEGDME and CO_2 aceton presented as well as a resorber-degasser heat ex-

change introduced as cycle modification

65 DeGroot, S. R., Thermodynamics of Irreversible Processes, Chapter VIII, p. 141 - 159, NORTH-HOLLAND PUBLISHING COMPANY. 1959

66 Habermann and Stetefeld, Bericht über eine ausgeführte Absorptionskälteanlage, Zeitschrift für die gesamte Kälteindustrie, 12, p. 121. 1905

67 Hausen, H., Wärmeübertragung im Gegenstrom, Gleichstrom und Kreuzstrom, Springer, Berlin-Göttingen-Heidelberg. 1950

68 Hellmann, H.-M., Chr. Schweigler and F. Ziegler: A simple method for modeling the operating characteristics of absorption chillers, Séminaire EUROTHERM N° 59, 6-7 Juillet 1998, p. 219 - 226. 1998

69 Altenkirch, W., assets of Edmund Altenkirch, 2003, archive of Wolfgang and Dr. Christel Altenkirch, Neuenhagen near Berlin, Bahnhofstr. 15. Commentary: Two volumes of letters by Altenkirch, (typewritten and collocated by Eva Hermann, née Altenkirch (daughter)), separated into private and business correspondence

70 Hoffmann, H., Experimentelle Studien über die Nutzeffekte von Thermoketten, Diss. Rostock, 1898, at Altenkirch, W.

71 Joffe, A. F. et al., Grundlagen der thermoelektrischen Kälteerzeugung (translated), Cholodilnaja Technika (Russian), Volume 33 (1956) No. 3, pp. 5 - 16. 1956

72 Justi, E., Leitungseigenschaften und Leitungsmechanismus fester Stoffe, Göttingen, Vandenhoeck & Ruprecht 1948 as well as in Kältetechnik Volume 5 (1953) p. 150. 1948

73 Justi, E., Elektrothermische Kälteerzeugung, Kältetechnik, Volume 5, p. 150. 1953

74 Justi, E., Elektrothermische Kühlung und Heizung. Grundlagen und Möglichkeiten, Arbeitsgemeinschaft für Forschung des Landes Nordrhein-Westfalen, Köln & Opladen, 70, Westdeutscher Verlag. 1958

75 Koch, B., Wärmeinhaltsdiagramm für Holz-Wasser, überlagert von einem T-logx-Diagramm für feuchte Luft mit relativer Feuchte als Parameter, Wärme- und Kältetechnik, in [26], p. 29. 1933

76 Kollert, J., Über den Wirkungsgrad der Thermosäulen, E.T.Z. 11, pp. 333 - 339. 1890

77 Kollert, J., Untersuchungen über die Verwendbarkeit der Thermosäulen für den elektrischen Großbetrieb, Zwölfter Jahresbericht (1889/92) der Naturwissenschaftlichen Gesellschaft zu Chemnitz, Volume 10, 16. 1893

78 Lorenz, H., Die Ermittlung der Grenzwerte der thermodynamischen Energieumwandlung, Zeitschrift für die gesamte Kälteindustrie II / 1895—some-

times partly in Z. d. Ver. Dt. Ing. 1894, Beiträge zur Beurteilung der Kühlmaschinen, II. 1895

79 Maiuri, G., Neue Absorptionsmaschinen für sehr tiefe Temperaturen, Zeitschrift für die gesamte Kälteindustrie, 46, p. 169. 1939

80 Merkel, F. and Bosnjakovic, Diagramme und Tabellen zur Berechnung der Absorptions-Kältemaschinen, Springer. 1929

81 Mollier, H., Dampfdruck und Lösungswärmen der Ammoniak-Wasser-Lösungen, Mitteilungen über Forschungsarbeiten auf dem Gebiete des Ingenieurwesens, insbesondere aus den Laboratorien der technischen Hochschulen., No. 63 and 64, Springer 1907, 1909, also cf. Z. Ges. Kälteindustrie 16 (1909), p. 51

82 Munters, C. G., Entwicklung des Elektrolux-Kühlapparates, Zeitschrift für die gesamte Kälteindustrie, Volume 38, pp. 197 - 200 and pp. 216 - 220. 1932

83 Nesselmann, K., Tiefste erreichbare Grenztemperaturen bei Absorptionsmaschinen mit Gasumlauf und Ammoniaklösung, Zeitschrift für die gesamte Kälteindustrie, Volume 35, pp. 197 - 221. 1928

84 Nesselmann, K., Zur Theorie der Wärmetransformation, Volume 12, No. 2, p. 89, paper Siemens-Werke. 1933

85 Nesselmann, K., Ein Übersichtsdiagramm für das Verhalten von Absorptionsmaschinen mit druckausgleichendem Gas, Zeitschrift für die gesamte Kälteindustrie, Volume 40, p. 117. 1933

86 Nesselmann, K., Die Verwendungsmöglichkeiten von Absorptionsmaschinen zur Wärmekrafterzeugung, Zeitschrift für die gesamte Kälteindustrie, Volume 42, p. 8. 1935

87 Nesselmann, K., Theorie und Anwendungsgebiete des Kältevermehrers, Zeitschrift für die gesamte Kälteindustrie, Volume 42, p. 213. 1935

88 Nesselmann, K., Der Verwendungsbereich von Absorptions-Kältemaschinen in Anlagen mit Bedarf an Kälte, Wärme und mechanischer Energie, Kältetechnik, Volume 1, p. 50. 1949

89 Nesselmann, K., Allgemeine Wärmetechnik / Kältetechnik, Volume 4 No. 12 /Volume 5, p. 252 / No. 12. 1953. Commentary: Efficiency in multistage designs for specific applications

90 Niebergall, W. and M. Gompertz, Untersuchungen an einer zweistufigen Ammoniak-Absorptionsmaschine—System Altenkirch, Zeitschrift für die gesamte Kälteindustrie, Volume 39 (1932), pp. 69, 140, 158, 187, 205, 222. Amendment in Volume 40 (1933) p. 13. 1933

91 Niebergall, W., Einstufige Absorptionskälteanlage für -45 °C Verdampfungstemperatur, Zeitschrift für die gesamte Kälteindustrie, Volume 43, No. 3, p. 51. 1936

92 Niebergall, W., Tiefkühl-Absorptionsanlagen mit zweistufiger Verdampfung, Zeitschrift für die gesamte Kälteindustrie, Volume 46, p. 81. Amendment p. 124. 1939

93 Niebergall, W., Arbeitsstoffpaare für Absorptionsmaschinen, Markewitz. 1949

94 Niebergall, W., Die Absorptionskälteanlage in der Brauereitechnik, Die Kälte, Volume 4, No. 9, pp. 237 - 244. 1951

95 Niebergall, W., Beitrag zur Geschichte der Absorptions-Kälteanlagen sowie Beitrag von Niebergall zu Plank, R., Handbuch der Kältetechnik, ID 103, Sonderdruck aus dem "Archiv für die gesamte Wärmetechnik" 1950 / aus der Allg. Wärmetechnik 1953 / and 1954, No. 7 and No. 8 / No. 2 and 3, pp. 139 - 143 and pp.169 - 175 / pp. 35 - 41 and 59 - 64 / pp. 33 - 39 and pp. 57 - 64. 1954

96 Pelletan, Thelen, Über die Entwicklung der Vakuumdestillation in "Beiträge zur Geschichte der Technik und Industrie", Jahrbuch 1 des Vereins deutscher Ingenieure, pp. 118 and 1909. 1834

97 Plank-Kuprianoff, Die Haushaltkältemaschine, III. Auflage, p. 274, Springer. 1948

98 Plank, R., Thermodynamische Untersuchung des Vorganges in der Absorptionskältemaschine aufgrund der Theorie der binären Gemische, Zeitschrift für die gesamte Kälteindustrie, Volume 17, No.1, p. 2. 1910

99 Plank, R., Neue amerikanische Absorptions-Kältemaschinen für Klima-Anlagen, Z. Ver. Dt. Ing. 91, pp. 493 - 496. 1949

100 Plank, R., Amerikanische Kältetechnik, 3. Bericht, Düsseldorf, pp. 77 et sequ., Springer, 1950

101 Plank, R., Erinnerung an Edmund Altenkirch, Kältetechnik, Volume 15, No. 12, p. 392. 1953

102 Plank, R., Hrsg., Verfahren der Kälteerzeugung und Grundlagen der Wärmeübertragung, Handbuch der Kältetechnik, Section B, I. Elektrothermische Kälteerzeugung, Volume 3, p. 62, Springer. 1959

103 Plank, R., Hrsg., Handbuch der Kältetechnik, Volume 3, p. 62, Springer-Verlag. 1959

104 Plank, R., Hrsg., Verfahren der Kälteerzeugung und Grundlagen der Wärmeübertragung, Handb. d. Kältetechnik, pp. 89 - 95, Springer-Verlag. 1962

105 Platen, B., von Munters, C. G., Om alstring av kyla (On generation of cold), Tekn. T55 (1925), pp. 89 - 95. 1925

106 Raicorla, C.,L.: An absorption refrigeration system may be your answer, Refrigerating Engineering., Volume 61, No. 3, pp. 276 - 279. 1953

107 Rayleigh Lord, On the thermodynamic efficiency of the thermopile., Phil. Mag., Volume. 20, pp. 361 - 363. 1885

108 Schilling, Edmund Altenkirch, Energietechnik, Volume 4, No. 3, p. 143. 1954

109 Schwarz, J., Transportkühlung von Lebensmitteln mit Wasser/Zeolith-Adsorptionssystem, 40, No. 6, p. 81, C. F. Müller, Heidelberg, la 11/94, 1994

110 Sprengel, Über die Anwendungsmöglichkeit der Kapillarkondensation in Adsorptionskältemaschinen. Papers from the company Siemens-Werke, Volume 20, No. 1, p.135. 1941

111 Stephan, K., D. Seher and R. Hengerer, Bau und Erprobung eines Absorptionswärmetransformators mit neuen Arbeitsstoffen, DKV Status Report No. 23

112 Stirlin, H., Die Verdienste Altenkirchs um die Absorptions-Kältetechnik, Kältetechnik, Volume 16, p. 9. 1981

113 Tchernev, D. I., Zeolites in Solar Energy & Refrigeration Applications—A Review, Ozone-Safe Cooling Conference, Washington.1993

114 Thomson, W., Proceedings of The Royal Society of Glasgow. 1852

115 Tou, D., Extended Irreversible Thermodynamics, Springer. 1993

116 Unger, S., Zur thermodynamischen Beurteilung der Verluste in Absorptionskälteanlagen, Die Technik, No. 11, Volume 9, pp. 609 - 613. 1954

117 Unger, S., "E. Altenkirch Grundgedanken und Ergebnisse seiner bedeutensten Arbeiten auf dem Gebiet der technischen Anwendungen der Thermodynamik", Die Technik, Volume 10, No. 2, pp. 113 et sequ. 1955

118 Unger, S., Über die Verbesserung des Wärmeverhältnisses an Ammoniak-Absorptionsmaschinen durch Anwendung wässriger Salzlösungen als Absorptionsmittel, Die Technik, Volume 12, No. 2, p. 119, Feb., 1957

119 Voigt, H., Kältemaschinenprozesse für Speicherbetrieb / Heat pumping and transforming processes with intrinsic storage, Die Kälte / Klimatechnik / Energy Convers Mgmt, Volume 35, No. 8 / Vol. 25 No. 3 / pp. 312 - 318 and 381 – 386. 1982

120 Voigt, H., Evaluation of Energy Processes through Entropy and Exergy, Int. Inst. For Applied Systems Analysis 2361 Laxenburg, Austria, Research memorandum

121 Voigt, H. and F.X. Eder, Ein neuer Arbeitsprozess für modifizierte Stirling-Maschinen—Der Stirling-Lorenz-Prozess, Vortrag, gehalten auf der Tagung der Dt. Physikalischen Gesellschaft, Berlin: March 30 through April 30,1987. 1987

122 Voigt, H., Peltier Cooling at Low Temperatures by Means of Semiconductors, Proc. of THE 10th INT. CONF. OF REFRIGERATION, Copenhagen, pp. 110-

114. 1959

123 Yazicilaroglu, S., Wassergewinnung aus der Luft, Zeitschrift für die gesamte Kälteindustrie, Volume 51, pp. 27 et sequ. 1944

124 Ziegler, F., Kompressions-Absorptions-Wärmepumpen., DKV Status Report No. 34, DKV. 1998

125 Ziegler, F., Sorptionswärmepumpen, DKV Status Report No. 57, Habilitationsschrift, DKV. 1997

126 Ziegler, F., Grossmann, G., Heat transfer enhancement by additives. Int. J. Refrig., Volume 19, pp. 301 – 309. 1996

127 Ziegler, F., Kahn, R., Summerer, F. and Alefeld, G., Multi-effect absorption chillers., TU-München, Physik-Dep., Institut E19. 1993

128 Ziegler, F., Kahn, R. Summerer, F. and Alefeld, G., Multi-effect absorption chillers, Physik-Dep., Institut E 19 der Technischen Universität München. 1987

129 Ziegler, F. and Alefeld, G., Coefficient of performance of multi-stage absorption cycles, Physik-Dep., Institut E 19 der Technischen Universität München, March 26. 1987

130 Ziegler, F. et al., Absorptionskaltwassersatz zur solaren Kühlung mit 10 kW Kälteleistung, Bayerisches Zentrum für Angewandte Energieforschung e.vom, ECOS * 98, pp. 573 – 579. 1998

131 Ziegler, F. et al., Absorptionskaltwassersatz zur solaren Kühlung mit 10 kW Kälteleistung. 2001

13.2   Patents

132 Altenkirch, E., Zentralheizung mit Verwendung der von einer Kältemaschine (einer Kompressionsmaschine oder einer Absorptionsmaschine) stammenden Abwärme, D.R.P. 249 915 vom 08. 01. 1911 / D.R.P. 249 916 vom 08. 01. 1911

133 Altenkirch, E., Thermosäule, D.R.P. 291 521 vom 04. 03. 1915, 1915. Commentary: Application for protection of the thermocouple Sb2Te3 against bismuth with 1/10 antimony

134 Altenkirch, E., Verfahren zur Nutzbarmachung der durch die adiabatische Kompression erzeugten Wärme höherer Temperatur bei Kompressionsmaschinen, D.R.P. 341 457 vom 14. 12. 1919, 1919. Commentary: Separated heating up to a higher temperature of a small amount of cooling water

135 Altenkirch, E., Verfahren zur Erzeugung von Warmwasser mit Hilfe von Kompressionskältemaschinen mit mehrstufiger Kompression und Zwischenkühlung, D.R.P. 338 283 vom 14. 12. 1919, 1919. Commentary: The intermediate cooling is enhanced to partial liquefaction, the intermediate condensate is supplied to the evaporator, the rest is further compressed

136 Altenkirch, E., Heizung, D.R.P. 330 378 vom 14. 12. 1919. Commentary: Step-wise warming-up of the utilizable medium with a heat pump, whereby the super-heating successively and finally the compression heat of the high-pressure condenser serve for the heating-up. Between the consecutive steps of the evaporators and condensers a heat exchanger is inserted.

137 Altenkirch, E., Absorptionsmaschine mit zwei Gefäßen verschiedenen Drucks, in deren einem absorbiert, in deren anderem entgast wird, D.R.P. 395 421 vom 10. 02. 1920, 1920. Commentary: Also see Section 6, Figure 6.1

138 Altenkirch, E., Verfahren zum Austauschen der Wärme von Flüssigkeiten oder flüssigkeitshaltigen Stoffen, D.R.P. 400 136 vom 16. 6. 1920. Commentary: Heat exchange due to evaporation and condensation, counterflow-wise due to separation into temperature ranges with the maintenance of the difference of the pressures

139 Altenkirch, E., Kälteerzeugungsanlage, D.R.P. 396 878 vom 08. 11. 1921, 1921. Commentary: Equalization of the pressure fluctuations with the compression due to the combination of piston and steam jet compression by means of a gas buffer.

140 Altenkirch, E., Absorptionsmaschine, D.R.P. 427 278 vom 17. 06. 1922, 1922. Commentary: The cryothermal apparatus driven by a gas bubble pump, see Section 6, Figure 6.2.

141 Altenkirch, E., Kältemaschine aus nur einem Material, D.R.P. 454 564 vom 07. 03. 1923, 1923

142 Altenkirch, E., Verfahren zur Aufrechterhaltung des Partialdruckes eines Dampfes in einem Gasgemisch, D.R.P. 388 717 vom 19. 01. 1923, 1923

143 Altenkirch, E., Verfahren und Vorrichtung zum Auskristallisieren von in Wasser gelösten Stoffen, D.R.P. 496 214 vom 08. 04. 1923, 1923

144 Altenkirch, E., Vorrichtung zur Entlüftung der Apparatur bei der Raffination mit flüssiger schwefliger Säure, D.R.P. 413 156 vom 15. 02. 1924, 1924

145 Altenkirch, E., Verfahren zur Wärme- oder Kälteerzeugung oder zur Wassergewinnung oder Trocknung mit Hilfe atmosphärischer Luft, D.R.P. 528 691 vom 06. 09. 1929, 1929

146 Altenkirch, E., Verfahren und Vorrichtung zur Befeuchtung und Entfeuchtung von Luft, D.R.P. 628 095 vom 23. 02. 1933, 1933. Commentary: Periodically regulated flow direction due to the alteration of the heat supply / dissipation of the perfused apparatus with surfaces according to D.R.P. 692 095, and with intermediate condensation.

147 Altenkirch, E., Absorptionskälteapparat, D.R.P. 614 397 vom 01. 03. 1933, 1933. Commentary: Similarity due to the patent D.R.P. 628 095 on Feb. 23, 1933 with an apparatus periodically operating due to carrier gas and with a changing flow direction caused by the alteration of the heating (two genera-

tors-absorbers)

148 Altenkirch, E., Verfahren und Einrichtung zur Trocknung von feuchten Stoffen durch Luft, D.R.P. 637 167 vom 11. 03. 1933, 1933. Commentary: Cyclic movement of hygroscopic materials in counterflow direction to the air

149 Altenkirch, E., Verfahren zum Betriebe einer Absorptionskältemaschine, D.R.P. 609 104 vom 20. 06. 1933; 1933. Commentary: Methods according to patent D.R.P. 637 167 with a regulation due to the position of the sun

150 Altenkirch, E., Verfahren und Vorrichtung zur Wassergewinnung aus der atmosphärischen Luft, z.B. mit einem geneigten Sonnenkollektor mit Glasplatte, D.R.P. 663 920 vom 29. 09. 1933, 1933

151 Altenkirch, E., Zum Trocknen und Kühlen dienende Vorrichtung, D.R.P. 692 693 vom 03. 10. 1935, 1935. Commentary: These are walls which have on one side a hygroscopic material as a coating, which can be easily perfused by water vapor, but on the opposite side it is impermeable for water vapor, for example due to an applied metalization

152 Altenkirch, E., Vorrichtung zum Eindicken von Flüssigkeiten, zum Kühlen von Luft oder Wasser, D.R.P. 715 871 vom 27. 10. 1936. Commentary: Water removal is caused by a wall permeable for water in relation to the content of the liquid of a wall with a big surface. On the other side of the wall a gas of high dryness flows, for example carbon acid.

153 Altenkirch, E., Verfahren zur Trocknung der Luft in feuchten Räumen, D.R.P. 664 953 vom 30. 12. 1936, 1936. Commentary: Specialization in heating due to combustion heat

154 Altenkirch, E., Verfahren zum Betrieb von Wärmeregeneratoren, D.R.P. 715 568 vom 07. 01. 1937. Commentary: Specialization in heating due to combustion heat

155 Altenkirch, E. and W. Niebergall, Verfahren und Vorrichtung zum Heben der einem Wärmeträger entzogenen Wärmemenge niedriger Temperatur auf eine höhere Temperatur, D.R.P. 867 122 vom 28. 08. 1950, 1950. Commentary: Use of several compression stages, in order to be able to create different temperature bands with a big degasification width in the sorption and desorption vessels of a compression machine with solution cycle (e. g. a heat pump)

156 Altenkirch, E., Unger, S., Verfahren zur Trocknung komprimierter Luft, D.R.P. 849 740 vom 25. 07. 1950, 1950. Commentary: Utilization of the compression heat for the regeneration of the hygroscopic materials which serve for the water removal

157 Altenkirch, E. and W. Niebergall, Verfahren und Vorrichtung zum Betrieb einer Wärmepumpe, D.R.P. 953 378 vom 29.8.1950, Commentary: Method analogously to the method which is described at Section 9.1 for the utilization of the outdoor cold

158 Altenkirch, E., Kompressions-Kälteanlage, D.R.P. 830 801 vom 25. 07. 1950, 1950. Commentary: Generation of cold or heat by means of several compression stages by aspiration of the vapors due to separately arranged evaporators in the direction of the to be cooled or heated media (implementation of the Lorenz processes)

159 Altenkirch, E., Verflüssiger für Kältemaschinen, D.R.P. 831 257 vom 13. 9. 1950. Commentary: Increase of the economical feasibility total of facilities for air cooling and air recooling with an uneconomical high air temperature due to the temporary use of a pure water cooling at an optimal outlet temperature of the cooling water

160 Altenkirch, E., Absorptionsmaschine zur kontinuierlichen Erzeugung von Kälte und Wärme oder auch von Arbeit, D.R.P. 278 076 vom 12. 08. 1911, 1911

161 Förster, H., Absorptionskälteverfahren für Temperaturen unter 0 °C, DE 102 19 262, B4 2004 09 16 vom 30. 04. 2002, 2004,

162 Geppert, H., D.R.P. 122948, 1899

163 Unger, S., Verfahren und Vorrichtung zur Förderung der Lösung in einem Absorptions-Kälteapparat. WP 17a / 61 742 vom 12. 06. 1959, 1959, Patentschrift Nr. 47417

164 Unger, S., Verfahren zum Betriebe einer mit Gasblasen-Förderung kontinuierlich arbeitenden Absorptionsmaschine, Auslegeschrift 1 153 038 vom 22. 08. 1963, 40, No. 6, p. 81. Commentary: Characteristic features are, among others:
   • A contraption to prevent the freezing of the condensate due to the pulsating supply of the condensate to the evaporator.
   • A big diffusion surface for the material transmission during the absorption due to the application of a film absorber with the trickling-down of the solution.

165 Voigt, H., Thermoelektrische Anordnung, Auslegeschrift Nr. 1 165 114 vom 28. 12. 1962 / Patentschrift Nr. 1 163 415 vom 28. 12. 1962

166 Voigt, H., Thermoelektrische Anordnung, Patentschrift Nr. 1 163 415 vom 28.12.1962, Auslegeschrift Nr. 1 165 114 vom 28. 12.1962

167 Zeuner, G., Technische Thermodynamik, Patentschrift Nr. 1 163 415 vom 28.12.1962, 1887

13.3 Miscellaneous

168 Lorenz, H., Lehrbuch der techn. Physik, 1904 (Hrsg.), Author by the Volume II: Technische Wärmelehre, Technische Wärmelehre, II, 1904

169 Lorenz, H., Zeitschrift f.d. ges. Kälteindustrie, 1894 (ab 1909 ist Hrsg. d. Dt. Kälteverein), Zeitschrift f.d. ges. Kälteindustrie, 1909, Felix, Leipzig

170 Linde, C. von, Kältetechnik, Beitrag zu Luegers Lexikon der ges. Technik, Aufl.

II (Z.f.d.ges. Kälte-Industrie), Volume V, ab S. 268, 1894, Lorenz, Hans, ab 1909 Dt. Kälteverein

171 Erdmann, E., née Altenkirch (daughter), In Search of Values for Human Survival, 1987, Columbia Pacific University. Dissertation for the attainment of the academical degree Doctor of Philosophy (Ph.D.) [translator's note]

172 Kruse, H., Grundlagen der Kälte- und Wärmepumpentechnik—Teil 2, Institut für Kältetechnik und Angewandte Wärmetechnik, Universität Hannover, 1982

173 VDI-Nachrichten, Düsseldorf, 30.10.09, wop

174 Erdmann, Erika. Forging a Human Future. Edited, and with an Introduction and Notes, by David Stover. Special Edition of Humankind Advancing Vol. 17, No. 1 & 2, Winter 2012/13. Rocks Mills Press. 2012. Canada. [translator's note]

175 Lötstelle eines Thermoelements—themojunction [English translation]. From Technik-Wörterbuch englisch-deutsch, deutsch-englisch—Messen • Steuern • Regeln. Dr.-Ing. Hans Dieter Junge. Verlag Technik Berlin 1972, p. 200 [translator's note]

176 Hermanns, Hubert. Techno-Diktionär. Deutsch-Englisch – English-German, Dritte erweiterte Auflage. Verlag Wilhelm Knapp, Halle (Saale) 1950. Third Enlarged Edition. [translator's note]

177 Meyer, Hans Joachim and Hans Heidenreich. English for Scientists, A Practical Writing Course. 1. Auflage. Verlag Enzyklopädie. Leipzig. 1990, First Edition. [translator's note]

178 2009 ASHRAE Handbook—Fundamentals (SI), Chapter 37, Abbreviations and Symbols [translator's note]

Note: The abbreviation Z. VDI in note 32 refers to *Zeitschrift Verein Deutscher Ingenieure*, the *Journal of the Association of German Engineers*.

14 Picture Credits

The Figures were obtained, after revision, edition, and computerized image generation by Jörn Schwarz, the co-author of the book, from Altenkirch's articles in the periodical "Zeitschrift für die gesamte Kälteindustrie" and from his published monographs in the publishing company "Verlag Technik" of the GDR, for example from [66], [40]. For Figure 6.6 we would like to thank Prof. Horst Kruse.

15 Acknowledgments of the Transcriber

First and foremost I would like to thank my father, *Siegfried Unger*, for implementing this project. Closely linked to the subject is the life of my parental family. My father met Altenkirch in Zwickau and followed him to Neuenhagen near Berlin. He was fascinated by his charisma as an inventor, researcher, and natural scientist. Altenkirch had deep roots in the Berlin scientific landscape; he was a pupil of Max Planck and had exchange of letters with Albert Einstein.

My father's life and thus the parental family of mine was closely related, coined, and interwoven in these early days with my father's profession as a close collaborator of, and later close friend to Altenkirch and his family. I still remember Altenkirch's gestalt and his left artificial arm, and his wife and children sitting around the big table in the living room, or picking fruit in the big orchard of his estate in Neuenhagen, in those times wonderful pears. After Altenkirch's death we often visited his wife, Margarete Altenkirch and continue to have contact with their descendants.

Secondly, I would like to thank **David Stover**, the Canadian successor and co-author of Erika Erdmann, Altenkirch's eldest late daughter. She emigrated from East-Germany to Canada in 1953.

David Stover, as the first native speaker, took kindly over the editorship of the whole manuscript of this transcription, and contributed a lot of his knowledge in his excellent journalistic style with rewriting the text while maintaining the original tone. He was the president and vice president of the Oxford Press Canada over several years, is a journalist, and an author (he was several times a co-author to Erika Erdmann) and also the successive editor and publisher of the journal "Humankind Advancing," which Erika Erdmann founded in Canada, and which she edited and published until close to her death in 2006. With the latest special edition of this periodical Humankind Advancing *Forging a Human Future*[174], published in the publishing company **Rock's Mills Press**[111] of which David Stover is the head, which comprises a content of 120 pages, he shows Erika Erdmann's life and work in a concise form. With Erika Erdmann's scientific and literary work of this periodical but also of her manifold publications as papers and also in book form, including her bachelor's, master's and doctoral theses, which were the basis of and surrounded this periodical, her striving for a peaceful world and the survival of mankind and its humanity, embedded in both the preserving of the living space of the human species and its ecological and social environment, was her interest. Starting from constructive discussions with her father, Edmund Altenkirch, on ecological issues and the development of her concern for the threat to nature, early on when she still lived in Neuenhagen near Berlin in East-Germany, led to her lifelong research in peace and futurology and thus to her numerous publications, with which she and her father were already far ahead of its time.

David Stover laid the basis for this transcription.

In the year 2001, due to my interest in linguistics I joined Connect—The English Speakers Cultural Club—Berlin, an English language club which filled, in those times, a market gap for the 60% or so of the Berlin German population who understand and speak English and wished to practice. This club brought together native English speakers, Germans who had lived abroad and others who simply loved to speak English.

---

[111] in which publishing company also this transcription is published [translator's note].

Concomitantly, the Berlin Friends Club, also an English Speakers Club, was founded which I have been attending since 2010. There, I met **James Hobson**, a Londoner, with whom I had interesting talks about engineering physics, the subject this transcription deals with. The mediation of *Gabriele Wandelt-Gärtner*, also a member of the latter club, initiated a collaboration with *James Hobson*, which was so fruitful that the rest of the text of the transcription could be assessed and proofread by him; for this extraordinary work I have to give him my warmest thanks. When I asked him once to write something about himself, he put it that way: "As for me personally, I'm not sure what I should write, I have no publications to my name, and nothing really relevant to write about. I was born on 24 June 1976 in London, I read 'Special Honours Chemistry with Analytical Chemistry' at The University of Hull between 1994-1998 and graduated with a 2:1. I have lived and worked in Berlin for 3 years and am currently engaged in delivering ecommerce solutions. I am a native English speaker, with a shaky understanding of German, and an even worse grasp of Russian. ..."

James Hobson continued and finalized the proofreading.

Furthermore I would like to thank my wife, Marianne Unger, née Papendorf, and my children, Henry Papendorf and Matthias Unger, as well as my sister, Verena Kanitz, née Unger, for promoting this work with their patience.

The whole might not have happened if I had not had such a lot of impact by my English-speaking friends, be it expatriates, or Germans who had been living abroad, or just people who would like to learn and speak English, who I met in these Berlin English Clubs.

My thanks go especially to the hard core of the English-speaking language club I first encountered, "Connect—The English Speakers Cultural Club—Berlin e. V."

Special thanks go to Philip Bacon, also a member of the latter club, whom I would like to thank for special formulations, critical statements, and literature suggestion.

I had constructive discussions with members of the Berlin Friends Club, such as James Hobson, and Junaid Shaikh, about the subject of this book, whom I also would like to thank very much.

Especially thankful I am to Ilse Sonntag, an English teacher, with whom I played the viola at the same stand in the Orchester Äskulap Berlin e. V., for her constructive advice in English Grammar, and revising the manuscript.

Furthermore I would like to thank my mother, Christa-Maria Unger, née Popp. Without her picking an announcement from a well-known Berlin newspaper about a freshly founded English language club (Connect Berlin) in the year 2001, with the header "Requirements are high," everything would not had happened. What followed was my successive regular attendance of this club with the respective consequence of preoccupying myself with the English language.

Related with Erika Erdmann's scientific work, and due to family reasons, I highly appreciate Erika Erdmann's concern for the threat to nature and I made it my own interest by studying her publications; last but not least, influenced by the scientific education of my father, and the interest in the English language of my father's sister, Annemarie Zipfel, née Unger, I pursued my father's intention to transcribe his German publication: "Edmund Altenkirch—Pionier der Kältetechnik."

Berlin, in March 2019, Michael Unger

16 Personal Afterword by Siegfried Unger

On the temporary relocation Altenkirch's to Zwickau

At the beginning of the year 1946 a refrigeration engineer named Köhler—proprietor of the company "Kälte-Köhler" in Zwickau (Saxony)—called on Edmund Altenkirch in Neuenhagen near Berlin, to call him, on behalf of the Soviet Military Administration in Germany (SMAD), to Zwickau [Saxony] for taking over a research contract in his company. It dealt with the development of a mobile cold storage system. Further assignments still ensued. For their processing, Altenkirch called, besides the former docents Osterland and Riske of the Ingenieurschule Zwickau, also Dr. H. Dannies (Author of the "Lexikon der Kältetechnik")[112] for help to support him with the documentation of his ammonia resorption chryothermal apparatus (for the earlier development of this machine, see Section 6) Afterward, Dr. Dannies appeared in the Zwickau company.

In addition to that, Altenkirch was entrusted with the management of a project for the climate control in cold-storage rooms. For this, he employed about 15 collaborators. The owner of the company, Köhler, and his wife were at this time, after eviction due to the SMAD, subtenants in my parent's apartment. Due to Mister Köhler's mediation I completed a job interview with Altenkirch and became his collaborator on this project.

The pieces of work for the SMAD were finalized at the end of the year 1948.

In Altenkirch's Neuenhagen office/laboratory there was (as of 1949) lots of work, inter alia, on a set of tables for the "delaying function"[24] which Altenkirch developed in connection with his research on the "compression machine with solution cycle"[19]. Furthermore, there had to be processed comprehensive mathematical calculations of curve shapes as well as graphics in the framework of the elaboration of manuscripts for Altenkirch's intended works for the publication in book form. In support of this work, Hans Voigt (brother of my schoolfriend Gerhard from Zwickau) was employed.

For the processing of the construction drawings, which had to be included into the monographs, the engineer Hans Lippschütz, later the joint partner of the GfKK[113], was hired from the Kälteinstitut Berlin which was in process of getting established. Designated head was Dr. Bock who looked for a contact to Altenkirch and his key research areas. In the year 1947 the Doctor of Engineering, Franz Xaver Eder, tenured Professor and Director at the "III. Physikalischen Institut"[114]—which dedicated itself due to him to the generation of low temperatures, especially for the research of structural properties of matter, especially for metals—established the contact to Altenkirch.

The result was Prof. Eder's involvement in the "Arbeitsausschuss für Kältetechnik"[115] of the "Kammer der Technik der DDR"[116]. After Altenkirch's death in the year 1953, his re-

---

[112] Dictionary of the Refrigeration Technology [translator's note].

[113] Gesellschaft für Kältetechnik-Klimatechnik mbH, Köln (Society for Refrigeration and Climate Technology, Cologne [translator's note]).

[114] Third Department of Physics of the Berlin Humboldt University [translator's note].

[115] Working Committee for the Refrigeration Technology [translator's note].

[116] Chamber of Technology, GDR [translator's note].

search establishment was continued, under the name "Laboratory Dr. Altenkirch" as a branch of the mentioned Third Department of Physics of the Berlin Humboldt University, with my leadership until the end of the ongoing work in the year 1956.

The topics of the diploma theses of Hans Voigt's and mine were attuned with Altenkirch and were included in his thematic areas (developments on low-pressure cryothermal apparatuses, power generation from heat sources with low driving temperature differences (waste heat) due to "investigations of adiabatic flows of a boiling liquid in cylindrical pipes" (1958), and "vapor pressure measurements on the ternary mixture: ammonia/water/calcium chlorine" (1956), to the latter also see [72]).

Altenkirch was, save those which were already mentioned here, also the teacher of the professors Nesselmann and Niebergall, who were known in the public circles of the German refrigeration technology, as well as Bertolt Koch, editor of the periodical "Allgemeine Wärmetechnik"[117].

After Altenkirch's death, the monograph "Absorption Refrigeration Machines" were brought by Hans Voigt and I to the publisher "VEB Verlag Technik der DDR". In the year 1961 Prof. Eder followed a call to the Walther-Meißner Institute of the Bavarian Academy of sience as a counsel of the research laboratories of the company Siemens AG. Hans Voigt followed Prof. Franz Xaver Eder (see further below) to Bavaria where he was at last Deputy Director for Energy Systems of the company Siemens in Erlangen.

After finalizing the work in Altenkirch's research establishment in Neuenhagen, I changed to the newly founded Institute for Cybernetics and Information Processes (ZKI) of the Academy of Science of the GDR.

After marriage to the music student Christa-Maria Popp in our mutual home town Zwickau in the year 1950, we had already moved from Zwickau to Neuenhagen at the beginning of Altenkirch's work after his return to his Neuenhagen residence where I, following the call of Altenkirch's who already had started with the work in his Neuenhagen research establishment in the year 1949.

Posthumously, we are indebted to Margarethe Altenkirch since she integrated us kindly into her big family (6 children, residing in their parent's apartment, or in the vicinity of it) and gave us solid support and encouragement. Her eldest daughter advocated the human advancement of mankind in North America[118].

I thank my wife for her support she provided as a whole under the complicated post-war conditions and the concomitant purpose of my studies at the Humboldt University with low earnings for Altenkirch and the scholarship, and the births in the years 1952 and 1955.

---

[117] General Heat Technology [translator's note].

[118] On the suggestion of Robert Muller, she established the "Global Association of Scientists and Thinkers Concerned with Long-Range Evolution", was founder, editor, and publisher of a quarterly entitled "Humankind Advancing"[59],and moreover published, inter alia, books on issues of "Realism and Human Values".

Her general interest she formulated with the words, "[My quarterly's] intend, to attract attention to the work of persons with the gift to lead our species to a higher stage of mental maturity...

Again and again I was stunned by the amount of time and energy expended by the most enlightened intellectuals to downplay one another's contributions—while the desperate global situation (each one seriously intends to improve) needs the combined impact of all their work."... (original English quote: Humankind Advancing, Vol. 14, No. 1 January 2003, p. 3 [translator's note]).

17 Index